AF371505

Reactive Species:

Manipulations Toward Desired Outcomes in Biological Systems

Reactive Species:
Manipulations Toward Desired Outcomes in Biological Systems

G K Suraishkumar

Indian Institute of Technology Madras, India

NEW JERSEY · LONDON · SINGAPORE · BEIJING · SHANGHAI · HONG KONG · TAIPEI · CHENNAI · TOKYO

Published by

World Scientific Publishing Co. Pte. Ltd.

5 Toh Tuck Link, Singapore 596224

USA office: 27 Warren Street, Suite 401-402, Hackensack, NJ 07601

UK office: 57 Shelton Street, Covent Garden, London WC2H 9HE

British Library Cataloguing-in-Publication Data
A catalogue record for this book is available from the British Library.

REACTIVE SPECIES: MANIPULATIONS TOWARD DESIRED OUTCOMES IN BIOLOGICAL SYSTEMS

ISBN 978-981-12-9073-2 (hardcover)
ISBN 978-981-12-9074-9 (ebook for institutions)
ISBN 978-981-12-9075-6 (ebook for individuals)

For any available supplementary material, please visit
https://www.worldscientific.com/worldscibooks/10.1142/13774#t=suppl

Typeset by Stallion Press
Email: enquiries@stallionpress.com

Preface

Reactive species (RS) play crucial roles in the cell to make life possible. What are RS? They are reactive molecules such as hydroxyl, superoxide, peroxide, and many others. RS are now accepted as the fundamental molecular mediators of stress effects on biological cells and have been implicated in many diseases, including cancer, cardiovascular, and neuro-degenerative diseases. The core thread in our research group has been RS and its manipulations for the past three decades. The systems to which they have been applied are over a wide range and topical — the system choice depended on the interest of students and funding agencies.

This book began as a research monograph. Later, I felt a need to provide some principles and background to help the reader. The initial four chapters are those principles and background developed from an elective course I teach at IIT Madras, Reactive Species in Medical and Related Technologies. Thus, this book can be used to teach a specialized course in RS as well. Chapters 5–10 cover our research journey over the past three decades in the area of RS manipulations in biological systems.

In Chapters 5–10, the aim was to narrate a story of our journey to maintain reader interest. Therefore, only indicative data are given in this book. The more complete data, mathematical models, explanations, and discussions are in our publications, to which references are provided at the appropriate places in the text. Further, the researchers who worked with me as students are referred to in the text by their names I am most familiar with. They are now, formally, Dr. Manjula Rao, Professor Susmita Sahoo, Professor Ganesh Sriram, and so on.

Reliving the past while writing this book was filled with gratitude to the Almighty and to my students, collaborators, funding agencies, and family for their dedication, involvement/support. It has indeed been a pleasure to look back at the contributions made in the past three decades. I feel content that we have made some meaningful and novel contributions to the field, some of which have already been applied in the industry for process improvements. I was lucky to settle down in this niche, novel area of manipulating RS within a couple of years into my independent research career, which began in 1993 at IIT Bombay. In addition, working with World Scientific has been pleasant — they allowed me four years to write the book because the work was still on, and I am thankful. In fact, the first paper from Chapter 10 on the RS module was published only a few months ago, after many years of work. Janani Venkatraman, our doctoral student with an artistic mindset, created an attractive, in-depth cover image that works at many levels for this book — I am thankful to her for the fitting cover image.

Chapters 5–10 of this book narrate our journey in RS thus far. For the remaining years until my superannuation (we have a hard stop at 65 in our system), I needed to pivot to specific aspects that would suit the many practical constraints. Thus, for the remaining research time, the choice is to look at RS manipulations for manufacturing biologicals in space/micro-gravity, a more applied aspect. Let us see what the future holds.

G. K. Suraishkumar
December 2023

Contents

Chapter 1

Relevance and Importance of Reactive Species

Reactive species (RS) are highly important intracellular molecules you may have never heard of. More specifically, they are the molecules to which you may have never paid attention. They have been known for many decades under different names — free radicals, reactive oxygen species (ROS), and now, more generally, RS. They are even part of the popular literature. For example, can you guess the source of the following excerpted passage?

> *What I've told you does have something to do with how we age. Because it appears, nature has a sense of humour. The thing that keeps us alive is also what causes us to age and eventually die. When oxygen reacts with our food in order to release energy, it also releases free radicals called oxidants. These oxidants are toxic. ...*
>
> *The oxygen helps convert the food we eat into energy. But, it also causes the release of oxidants into our body, which start reacting inside us. We rust from the inside and hence age and eventually die ...*
>
> *... the Somras, which when consumed, reacts with the oxidants and absorbs them and then expels them from the body as sweat or urine. Because of the Somras, there are no oxidants left in the body.*

If needed, please check the answer in the Additional Information section at the end of this chapter. RS play a prominent role in the plot, too.

1.1. Aging and Diseases

RS play essential roles in a biological cell, the fundamental functional unit of life, including cell signaling. However, they are notorious for their adverse effects. We first follow the popular view to establish interest and then take a more comprehensive view of the roles of RS.

Aging is of significant interest to humankind and has been so for millennia. A genetic disorder leads to *progeria*, which causes drastic premature aging. Children with this disorder do not usually live beyond their early teens. Please see additional information on the havoc that progeria causes in the lives of families with progeria patients, the causes at the molecular level, and possible cures through gene therapy approaches.

Now, let us ask ourselves, what is aging?

We are familiar with the external manifestations of aging in humans. Changes in skin properties, changes in facial and bodily appearances, loss of strength, loss of stamina, and so on, are well-known and observable signs of aging. With increasing age, there is a progressive loss in the functioning of the organs and tissues. Organs and tissues are made of cells, and we all know that the biological cell is the fundamental functional unit of life. So, let us ask ourselves, what happens at the cell level as it ages?

The following are known [DiLoreto and Murphy, 2015] to occur at the cellular level that accompany aging:

- Shortening of telomeres, the DNA sequences at the end of the chromosomes
- Improper regulation at the transcription level
- Improper nuclear trafficking
- Improper nuclear organization
- Improper protein translation
- Insufficient proteostasis, or maintenance of protein quality, including unfolded protein response
- Improper autophagy
- Improper maintenance of mitochondrial function, its biogenesis, and mitophagy (mitochondrial autophagy)
- Decreased cytoskeletal integrity
- Improper cell membrane and extracellular matrix

The accumulation of the above degenerative processes in the cell leads to damage to the cells, which progressively leads to cell dysfunction and, hence, failure of the organ comprising such cells.

When a cell ages, the RS, especially the ROS levels inside the cell, increase significantly. Some of the above cellular events, such as mitochondrial dysfunction and mitophagy, also contribute significantly to the increase in intracellular ROS. An increase in the highly reactive ROS increases oxidative damage to all the fundamental biomolecules inside the cell — nucleic acids, proteins, lipids, and carbohydrates. Thus, RS play a significant role in the cell's aging process.

Cells indeed divide, and new daughter cells result from the parent cell. Thus, the aging of the new daughter cell, which would be relevant only to its lifetime, may not be a concern. However, cells in different tissues divide at widely varying rates. For example, cells in the skin's top layer divide once every 30 days, white blood cells divide in about every two weeks, red blood cells divide in about 120 days, and liver cells divide in about 18 months. It is also known that differentiated neurons do not divide, but new neurons become available in certain parts of the brain from stem cells. Thus, some cells in the body, such as neurons, cardiac cells, and so on, live for a long time. In addition, even the cells that divide frequently can do so only a finite number of times (say, about 50 times, *in vitro*. Note: not all cells in an organ will divide simultaneously). Thus, the aging of cells could be a concern for their appropriate functioning and, therefore, a concern for the proper functioning of the tissues, organs, and the whole individual they comprise.

Now, let us look at diseases, some of which correlate with age. RS are implicated in many human diseases, including cancer, cardiovascular diseases, neurodegenerative diseases, and so on. We will see the details in a later chapter after we have acquired the necessary background on RS. Interestingly, RS act as a double-edged sword in cancer. At a certain level, RS cause cancer, whereas at much higher levels, they can kill cancerous cells. Cancer therapies such as chemotherapy and radiotherapy work on the principle of increasing RS levels inside the cancerous cells to kill them. Most chemotherapeutic drugs are redox agents, which increase the intracellular RS in cancerous cells.

1.2. Other Areas

RS are important in many areas of human endeavor. Some of those areas are listed below.

Food processing/preservation: RS are generated in many ways in food — reactions, both enzymatic and others, irradiation, and exposure to light all cause the formation of RS in food. The reactions of RS, especially ROS, with food molecules such as lipids, proteins, and carbohydrates cause undesirable compounds and even carcinogens. Food is usually a biological material and, hence, is fundamentally composed of biological cells.

Organic artifact preservation for museums: RS can cause significant damage to preserved artifacts. Special processes need to be employed to prevent RS damage in museum artifacts, especially organic artifacts.

Atmospheric pollution mitigation: It may be known that many reactions are simultaneously occurring in the atmosphere. The reactions also help neutralize some undesirable compounds emitted into the atmosphere by anthropogenic activities. Many of these reactions involve RS.

Polymerization (say, in plastics): RS play a central role in the production of plastics. Important reactions in the plastics production process are RS-based, as you may have learned in chemistry courses. Free-radical polymerization is a more common but older term used to denote the same.

Sonochemical processes: It is known that ultrasound significantly increases the rates of processes. The RS generated under the high-energy conditions of ultrasound-based cavitation seem responsible for the sono-chemical intensification of reaction rates.

Combustion: Combustion is important in many processes that have relevance in multiple contexts — energy generation, transport, pollution abatement, and so on. Many RS are involved in the extensive set of reactions in combustion.

1.3. What Exactly are Reactive Species?

The "definition" of RS is somewhat nebulous. It is more of an understanding that one becomes comfortable with experience rather than a clear-cut definition. In general, the RS are supposedly highly reactive molecules. However, there could be a drastically wide variation in their reaction rates. For example, the second-order rate constant of something called the Fenton reaction with Fe^{2+} (we will see the details in Chapter 2) is about 76 $M^{-1}\,s^{-1}$. In contrast, the second-order rate constant of some other reactions that involve the same RS as the products in the Fenton reaction, called hydroxyl radicals, could be about $10^{11}\,M^{-1}\,s^{-1}$.

The RS could be derived from a variety of molecules. The ones derived from oxygen are termed ROS. The term ROS is much more popular than RS because RS is a recent term. RS includes all RS, including the ones derived from nitrogen, termed reactive nitrogen species (RNS), the ones derived from sulfur, termed reactive sulfur species (RSS), the ones derived from halogens, termed reactive halogen species (RHS), and so on. Prominent examples of RS are presented in Table 1.1.

A closer look at the formulae of the most prominent RS given in the second column of Table 1.1 shows the presence of a dot in some, a negative sign in some, and both in some. What does a dot mean? What does a negative sign mean?

To understand the dot, let us recall the electronic structure of atoms that one would have studied in high school. We know that the electrons fill the orbitals in the order of their energy levels according to the Aufbau

Table 1.1.   Common examples of reactive species

Derived from oxygen	
Hydroxyl	$^{\bullet}OH$
Superoxide	$O_2^{\bullet-}$
…	
Derived from nitrogen	
Nitric oxide	$NO^{\bullet}$
Peroxynitrite	$ONOO^-$
…	
Derived from sulfur	
Thiyl	$RS^{\bullet}$
…	
Derived from …	

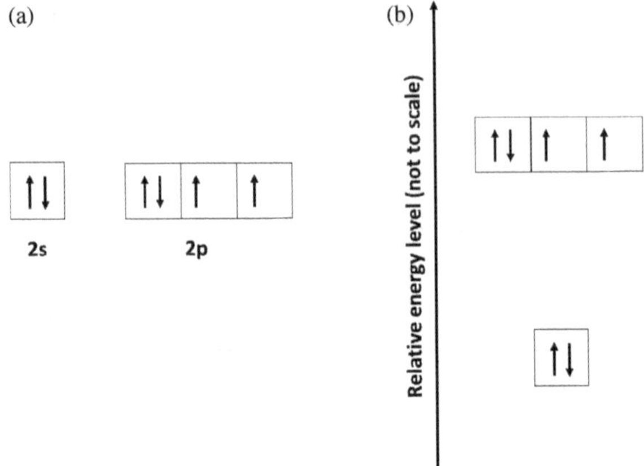

Figure 1.1.   The outer orbitals of the oxygen atom: (a) Standard representation. (b) The relative energy levels associated with the outer orbitals of oxygen

principle. Figure 1.1 shows the oxygen atom's outer orbitals, 2s and 2p. Recall that the atomic number of the O atom is 8; so it has 8 electrons. Two electrons are in the 1s orbital with opposite spins, two in 2s, and 4 are in the 2p orbitals. Two electrons in the 2p orbitals are unpaired.

Now, let us look at the outer (2s, 2p) orbitals of an oxygen molecule. Two oxygen atoms, say O (first) and O (second), constitute the oxygen molecule. The atomic orbitals of the two atoms overlap to produce the molecular orbitals. Figure 1.2 gives the relevant molecular orbitals in the order of increasing energies from the bottom to the top. Two electrons in the π^* (antibonding) orbitals are unpaired.

The dot represents the presence of one or more ***unpaired*** electrons in the relevant shells of the atomic or molecular orbitals.

Let us look at the molecular orbitals of some other prominent ROS in Figure 1.3. A species called singlet oxygen (one form of it) has two unpaired electrons; superoxide has one unpaired electron, whereas peroxide does not have any unpaired electrons. The spins of the unpaired electrons are opposite in this form of singlet oxygen, whereas they are the same in molecular oxygen. Superoxide and singlet oxygen have a dot in their formula that represents the unpaired electron(s).

On the other hand, a negative sign denotes something else altogether. It denotes a net negative charge on the atom/molecule. In other words, the

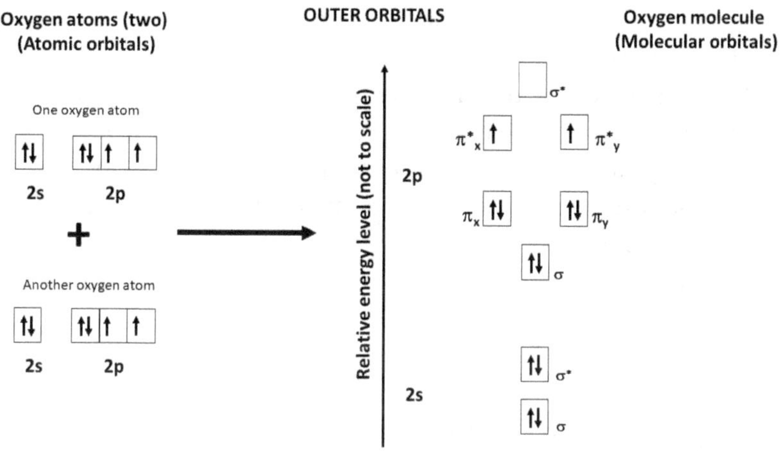

Figure 1.2. Atomic and molecular orbitals (outer) of oxygen atom and molecule, respectively

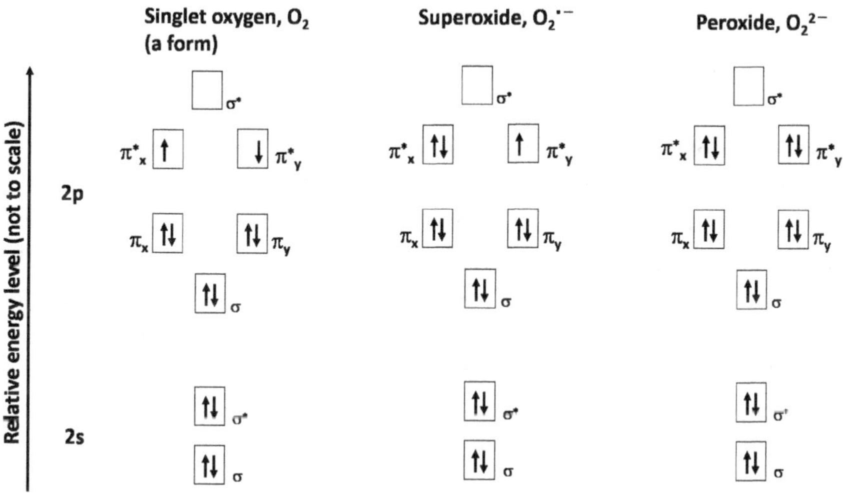

Figure 1.3. Molecular orbitals (outer) of some prominent reactive oxygen species (ROS)

number of electrons in the atom/molecule is more than the number of protons. The negative sign indicates that the molecule is ionic.

Not all RS need a dot or a negative sign — some RS do not possess unpaired electrons and are uncharged. A typical example is hydrogen peroxide, H_2O_2. Some may be charged but not have unpaired electrons; an example is peroxynitrite, $ONOO^-$.

The dot is not explicitly shown in the molecular formula for all molecules that possess unpaired electrons. A good example is oxygen, denoted by O_2. Although there are unpaired electrons in O_2, the dot is not explicitly shown while writing the molecular formula. Whereas they are always explicitly shown for superoxide and hydroxyl radicals. An inexperienced person in the field needs to become comfortable with this practice.

1.4. More on Oxygen

The RS derived from oxygen are the most popular, although the other RS also play important roles in a biological system. Interestingly, Earth had no oxygen when it was formed about 4.8 billion years ago. The first life forms did not have the metabolic machinery to process oxygen; they were anaerobes.

Cyanobacteria in the oceans are believed to be the predominant source of oxygen in the atmosphere. Cyanobacteria, by photosynthesis, could split H_2O into O_2 and H_2. The by-product of that process, O_2, accumulated in the atmosphere. The anaerobes do not have defence systems against the RS, especially the ROS. The ROS oxidize the fundamental biomolecules and render them nonfunctional. As the O_2 level increased in the atmosphere, the anaerobes, for survival, evolved with the development of cellular mechanisms to handle oxygen and became aerobes. Or, they moved to locations such as deep oceans or other places O_2 could not reach. The ones that could not evolve or move to oxygen-free locations became extinct. Links to relevant videos and figures are given in the Additional Information section.

The toxicity of oxygen to anaerobes is used to treat gangrene, for example. Gangrene refers to tissue death from the lack of blood circulation. A type of gangrene can be caused by infection with an anaerobic organism, *Clostridium*. One of the treatment methods, at least in part, is hyperbaric therapy, in which the affected part is exposed to high-pressure oxygen. The anaerobic organism cannot survive in the presence of oxygen, and thus, the severity of gangrene can be reduced.

Oxygen can also affect aerobic organisms/cells that have developed mechanisms to counter the ill effects of oxygen due to the generated ROS. Well-designed experiments reported in the literature (e.g., data on the effects of radiation on cell viability under different gas compositions from Michaels given in Halliwell and Gutteridge [1999]) have demonstrated the toxicity of ROS on mammalian cells. It is argued that the human body is designed to minimize the harmful effects of ROS — oxygen is 21% in air when it first enters the lungs. The oxygen partial pressure reduces as it reaches the different parts of the body through the bloodstream. It is argued that the reduction in oxygen partial pressure helps limit the generation of ROS and, thus, their deleterious effects on some body tissues/cells.

High oxygen levels are used in some therapies, such as treating COVID-19 ravages, nitrogen bends in divers who surface too quickly, and prematurely born babies. However, in such cases, it is a choice between two challenging aspects — the condition being treated and the ill effects of high oxygen levels. High oxygen levels cause a chronic toxic effect on the nervous system; convulsions can result. Also, pure oxygen can affect the respiratory system itself, designed to better handle the ill effects of atmospheric oxygen. For example, pure oxygen can cause cough, sore throat, and chest soreness in about 6 h. Alveoli damage can result in 24 h; that damage activates the body's repair mechanisms. The repair mechanisms lay down inelastic material in the lung (fibrosis), which permanently impairs gas exchange. Premature babies do not have well-developed defence systems against RS. Thus, even atmospheric oxygen levels of 21% by mole/volume can lead to the formation of a fibrous layer behind the eye lens (retinopathy of prematurity), resulting in eyesight loss. Thus, one needs to be careful in working with high oxygen levels.

1.5. An Important Set of ROS Reactions in an Aerobic Cell

Many reactions that involve RS, especially ROS, occur in the cell as a part of the metabolism, signaling, and so on. Let us see a fundamental set of ROS reactions here; we will see some others (only *some* others), as

$$O_2 + e^- \rightarrow O_2{}^{\bullet-}$$

$$O_2{}^{\bullet-} + e^- + 2H^+ \rightarrow H_2O_2$$

$$H_2O_2 + e^- + H^+ \rightarrow H_2O + {}^{\bullet}OH$$

$${}^{\bullet}OH + e^- + H^+ \rightarrow H_2O$$

Figure 1.4.   The conversion of oxygen to water by the addition of a total of four electrons — a four-electron reduction. This set of reactions occurs in every aerobic cell

necessary, as we progress through the book. The last chapter will consider many more reactions.

The set of reactions that describe the conversion of oxygen to water as a part of the fundamental metabolic processes in any aerobic cell is given in Figure 1.4.

The addition of one electron to oxygen results in superoxide.

The addition of two electrons to oxygen results in hydrogen peroxide.

The addition of three electrons to oxygen results in hydroxyl radical.

The addition of four electrons to oxygen results in water.

1.6. Antioxidants

We saw in an earlier section that when oxygen concentrations increased in the atmosphere as our planet evolved, some anaerobes developed defence mechanisms to tolerate oxygen. What are these defence systems? They are various ***antioxidant systems*** that counter the RS. In the context of our interest, an antioxidant is understood as a species that delays, blocks, or even mitigates oxidative damage to a biomolecule in focus.

Antioxidants could be enzymatic such as superoxide dismutase (SOD), which acts specifically against superoxide, or glutathione peroxidase, which acts against hydroxyl radicals. Some molecules, such as glutathione (GSH), a tripeptide, can act against various RS such as hydroxyl, hypochlorous acid, peroxynitrite, nitrogen dioxide, and others. Antioxidants could also be nonenzymatic molecules such as ascorbic acid or α-tocopherol, proteins such as transferrin that minimize the availability of prooxidants such as iron, or even proteins such as chaperones that protect DNA. Strategies such as compartmentalizing molecules (e.g., iron) to prevent their reaction with cellular H_2O_2 are also considered antioxidant strategies.

Some antioxidants, such as SOD and glutathione peroxidase, are synthesized in the cell, whereas others, such as ascorbic acid and α-tocopherol, are ingested through diet.

The net effect of RS generation and antioxidant action results in a certain level of RS at a particular time point in the cell. Since RS influence many critical functions in the cell (we will see some in the later chapters), there is a certain, desired, net-specific level of intracellular RS at any time. That desired level results from the balance between production and consumption rates of RS in the cell or RS *homeostasis*. The desired level or the homeostatic level differs with the kind of cell. For example, a cancer cell has a higher homeostatic RS level than a normal cell. In contrast, a stem cell has a much lower RS homeostatic level than a normal, differentiated cell. In addition, the desired level could vary with time — we discovered that rhythms exist in the pseudo-steady-state (PSS) levels of particular RS.

1.7. What is the PSS Level?

To understand the PSS level, let us first consider the assembly of a car engine. Let us assume that it takes about an hour to assemble an engine. Further, let us say that each bolt that goes into the engine assembly takes, on average, only about 5 s to make. Whether it takes 5 s, 3 s, or 10 s to make a bolt will not affect the engine assembly rate. It will still be assembled at a rate of 1 engine per hour, irrespective of the variation or the "unsteadiness" in the *rate* of bolt making.

In the above scenario, the engine assembly was the slower process, and bolt-making was the faster process. The difference between the rates of those two processes was significant — engine assembly was 1 h^{-1}, whereas bolt making was 5 s^{-1}. Our interest was in the slower process. If these conditions are met, then the unsteady nature of the much faster process does not affect the rate of the slower process. In such situations, the faster process can be assumed to be at a steady state compared to ***the slower process.*** This is the pseudo-steady-state approximation (PSSA).

Now, let us turn our attention to the system of our interest. Let us consider two processes: RS reactions inside the cell and cell growth. Let us say that our interest is in the process of cell growth. Typically, RS reaction rates are much higher than the cell growth rates. Thus, if our interest is in cell growth, the unsteady nature of the intracellular RS levels (variation in the intracellular RS levels with time in short enough time frames) does not affect the rate of cell growth, which is a much slower process. Therefore, we can assume the RS reactions to be at a steady state over reasonable time intervals while considering cell growth. Such steady-state RS levels are called PSS levels.

In an earlier section, we saw some ill effects of oxygen on even healthy humans and their cells. The ill effects manifest even though they have antioxidant defence systems. One of the common ways of explaining the ill effects despite the presence of cellular antioxidants is through the cell's capacity to handle a certain level of ROS. However, this widely prevalent view is limiting because the cell is a dynamic system — it is ever-changing with time. The widely prevalent view ignores the fact that there could be a continuous generation or degradation of ROS and antioxidants. The following section sheds more light on this aspect.

1.8. A Better Framework for Analyzing or Manipulating Biological Systems

The intracellular RS generated by cellular processes or otherwise are countered or quenched by the antioxidants that may be already present in the cell. In addition, more enzymatic antioxidants are generated through transcription and translation in response to the generated RS. It is important to realize that the intracellular RS generation and the RS quenching by antioxidants already present in the cell occur simultaneously. Many presentations tend to give the neat impression that RS are first generated and then quenched by the antioxidants in a sequential fashion. This impression or understanding is usually created because, from a **species view**, an RS **species** needs to be generated, and only then can that species be quenched. The impression is strengthened further by the discussion on the specific amounts (levels) or concentrations of RS. However, in a dynamic system such as a biological cell, the **processes** of generation and quenching occur simultaneously. Thus, the impression or understanding of sequential generation and quenching of RS is not useful when one is interested in manipulating biological systems, for example, in medical and production contexts. To illustrate, let us consider a measure in the medical context: the extent of the RS damage.

The RS damage extent that a few assays can measure can only provide the damage that has *already happened*, mostly *irreversibly*. The RS damage extent provides a *cumulative measure* of the RS effects, which has a *diagnostic* value. Treatment or *therapeutic* decisions are usually made to save the remaining parts of the tissue/cell, not to address the irreversible damage already done. Thus, a measure of the *current status* is needed for meaningful therapy decisions.

The cell is a dynamic system that constantly changes with time. For dynamic systems where multiple processes *simultaneously occur*, the rate of processes that occur in them is a much better framework/view/measure for meaningful understanding and manipulation. In this case, the *rate* at which the intracellular RS become available, the *rate* at which the antioxidants scavenge the RS, and if applicable, the PSS levels/concentrations in the system of interest, say, the cell, are much better useful than the measures that reinforce a species view (**cumulative amounts/concentrations of the species**).

This view has significant relevance in the use of antioxidants in disease contexts. The PSS values of specific intracellular superoxide (siSupOx) provide useful information on oxidative stress when the interest is a slower process, such as growth or even metabolism [Menon *et al.*, 2013]. Thus, if siSupOx levels are used to decide on an external antioxidant dosage, antioxidant therapies may be more effective. The current, widespread practice of using aspects/products of RS damage as markers of their levels could lead to overestimations of RS since these intracellular marker levels are cumulative. A measurement of these damaged products at a particular time point would correspond to amounts accumulated over a long time. It would not reflect the actual RS levels present at that time point. This is also a consequence of the enormous difference in rates; in this case, repair of damage/clearing away of damaged products/turnover is much slower process than reactions involving RS. Thus, if the antioxidant dosage is decided based on RS damage, it is inappropriate because it is based on a cumulative aspect and not on the status of the cells/tissue/organ when administering the antioxidant. For example, when the measured cumulative damage is high, the levels of RS may not be high. Thus, the unintended effects of the administered antioxidant could manifest, resulting in the strategy's failure. Such failures have often happened in the past, although, in principle, the antioxidant should directly reduce the RS level.

A similar consideration may be necessary to reduce the indiscriminate intake of antioxidants through diet. For example, suppose a time profile of RS in the relevant tissues can be generated. The antioxidant intake can be temporally designed to coincide with high RS levels to ensure effectiveness.

Therefore, it is recommended that due importance be given to the rates of RS formation and scavenging while making appropriate decisions for a biological system.

1.9. Oxidative Stress and its Effects on Cells

When the rate of RS production, especially ROS production, is higher than the rate of ROS neutralization by the various antioxidant mechanisms in the cell, the cell experiences ***oxidative stress***. Oxidative stress can have different, sometimes diametrically opposite effects on the cell, depending on its intensity.

Surprisingly, low levels of oxidative stress can increase the growth rate of cells and the rate of products made by the cell. The later part of this book extensively looks at this aspect.

The cell can adapt to oxidative stress under certain conditions by upregulating its counter mechanisms. Such upregulation can also make it robust to higher oxidative stress levels in the future.

As expected, when the cell experiences high oxidative stress levels, it causes damage to cellular molecules. Depending on the level of damage, the cell may be able to repair the damage later. However, such damage can result in senescence, the condition in which the cell is alive but cannot divide further.

At much higher levels of oxidative stress, the cell can die through programmed cell death (apoptosis) or necrosis and other related mechanisms.

1.10. The Reactive Species Level is Not Correlated with "Its" Corresponding Antioxidant Level in the Cell

It is commonly accepted that antioxidants, including enzymes such as SOD and nonenzymatic ones, interact with RS, such as superoxide (SupOx), to reduce the specific levels of intracellular RS. Further, the kinetic mechanism is known in many cases. For example, the kinetic mechanism for the reaction between SupOx and SOD [Bull and Fee, 1985], as well as the mechanistic structural details [Shin *et al.*, 2009], are known. There seems to be a widespread expectation that more SupOx leads to the formation of more SOD. The data from studies that have exposed *Escherichia coli* to high oxygen (*not* superoxide) concentrations or agents that generate SupOx, such as streptonigrin, paraquat, and pyocyanin, as well as similar studies on other bacteria, algae, animals, and plants, seem to support that expectation [Halliwell and Gutteridge, 2015].

Based on circumstantial evidence such as the above, antioxidant levels are commonly used as surrogate measures of the "cellular stress" caused by RS. Interestingly, no direct measurements of intracellular SupOx levels and corresponding SOD levels were reported before our paper [Kizhuveetil *et al.*, 2020]. The rates of reactions that involve RS are usually higher by many orders of magnitude than the rates of cellular or biological processes of interest, such as metabolism or growth. Thus, the concept of PSS of RS reactions can be effectively invoked for dynamic systems such as a living cell, by which the RS reactions can be considered to be at a steady state in comparison with the process of interest. Under such conditions, the PSSA allows the use of temporal RS levels as straightforward, effective oxidative stress markers. Further, it obviates the need to use surrogate markers such as antioxidant levels to measure oxidative stress. More importantly, as shown next, there seems to be no strong basis to use antioxidant levels to infer levels of the corresponding RS.

Counterintuitively, the results show no correlation between temporal levels of SupOx and "its" antioxidant, SOD.

The relationship between specific intracellular levels of SOD (siSOD) and specific intracellular PSS levels of superoxide (siSupOx) for three widely different systems, namely, SiHa cancer cells, *Chlorella vulgaris*, and *Bacillus subtilis*, respectively, under different conditions, are presented in Figures 1.5–1.7. SiHa cancer cells are mammalian cells, and *C. vulgaris* is a microalga; both belong to the eukaryotes domain, whereas *B. subtilis* is a bacterium, which belongs to the prokaryotes domain. Thus, the results seem valid over a wide range of biological systems. It is clear from the data in Figures 1.5–1.7 that counterintuitive to traditional expectations, there is no relationship between siSOD and siSupOx in any of the systems [Kizhuveetil *et al.*, 2020].

Further, the characteristic time constant (relaxation time) for transcription and translation processes is tens of minutes [Clark and Blanch, 1997], which can be a reasonable estimate for the characteristic time constants for synthesizing antioxidant enzymes. That characteristic time constant is many orders of magnitude higher than that involved in the reactions of SupOx — say 10^{-7} s; the relevant rate

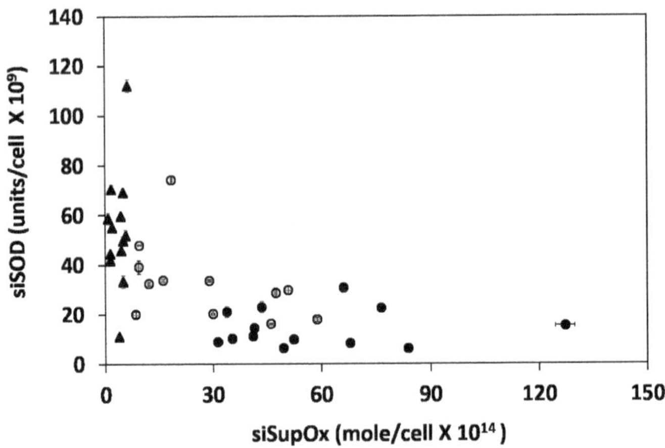

Figure 1.5. Relationship between specific intracellular levels of SOD (siSOD) and superoxide (siSupOx) in SiHa, a cervical cancer cell line (untreated: open circles; menadione treated: filled circles; curcumin treated: filled triangles)

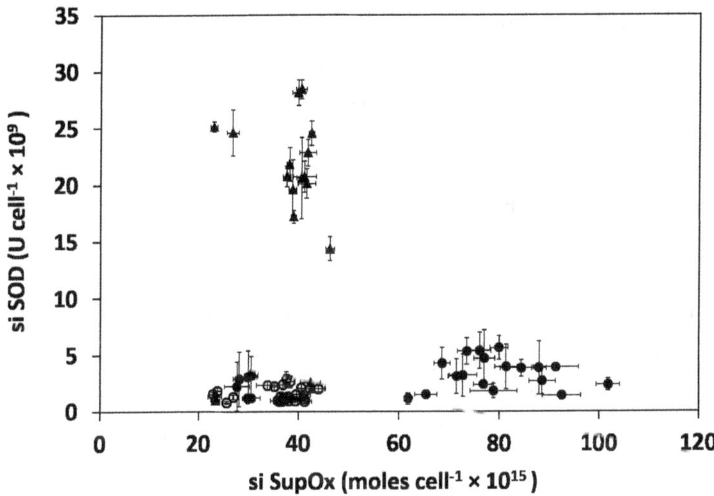

Figure 1.6. Relationship between specific intracellular levels of SOD (siSOD) and superoxide (siSupOx) in *Chlorella vulgaris* (untreated: open circles; menadione treated: filled circles; H_2O_2 treated: filled triangles)

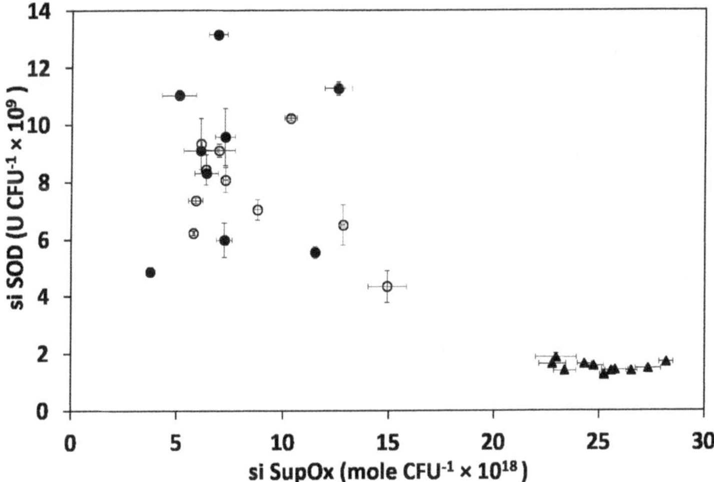

Figure 1.7. Relationship between specific intracellular levels of SOD (siSOD) and superoxide (siSupOx) in *Bacillus subtilis* (untreated: open circles; menadione treated: filled circles; H_2O_2 treated: filled triangles)

constants [Halliwell and Gutteridge, 2015] were used for the estimate. Since there is a significant mismatch in the rates of the two processes, it is unlikely that induction of the antioxidant enzyme alone, say SOD, by the RS, say SupOx, could be an effective defense against the deleterious effects of RS.

Another aspect is the interconversion between the various RS, say SupOx, hydroxyl, hydrogen peroxide, and others, at very high rates with second-order rate constants [Halliwell and Gutteridge, 2015] of the order of 10^8 M^{-1} s^{-1}. Thus, it may be inefficient for the cell to have a particular RS to induce its corresponding antioxidant enzyme. A pool of antioxidants is probably constitutively present in the cell to handle the relevant RS to minimize damage in both prokaryotes and eukaryotes.

The high rates of interconversion between the various RS do not undermine the use of RS level, say SupOx level, as an oxidative stress indicator. That is because we measure the net indicator concentration at any time, which results from the simultaneous processes that occur in the cell (the system) to create, consume, and/or transport the indicator at different, high rates. This fact, combined with using the PSS concept when the time scales involved are appropriately large, as mentioned earlier, makes RS level a suitable indicator of oxidative stress. It is similar to pH (or temperature) being a suitable indicator for a wide range of aspects, say, from the status of a reaction to that of a complex cell organelle, although many simultaneous processes could contribute to the net hydrogen ion concentration/activity at any particular time.

The levels of "total ROS" measured through, say, DCFH-DA, a popular fluorescent probe, may not be an appropriate measure of oxidative stress because the various ROS such as hydrogen peroxide, SupOx, hydroxyl, peroxyl radicals, and so on, have widely different damage potentials. The damage potential (an interpretation of the oxidative stress) indicated by a particular "total ROS" value could be different depending on the fraction of the above species at that particular time. For a similar reason, the "total antioxidant capacity" is unlikely to provide useful information because the effectiveness of different antioxidants against different RS are different. In any case, as demonstrated above, the individual antioxidant values are not surrogate measures even for "their" RS.

This first chapter provided some background information and principles to understand the importance of RS in biological systems. With this priming, let us delve into our RS journey through the following chapters.

Additional Information

The excerpt is from the chapter, "The Drink of the Gods," in the popular book:
Amish Tripathi, *The Immortals of Meluha*, Westland, 2010.

Video: Progeria Children — Rapid Aging Disease:
https://www.youtube.com/watch?v=AzyJyCZqumw

Video: Cell disruption caused by mutant prelamin: A protein points to origin of progeria:
https://www.youtube.com/watch?v=Mseec3-o6hY

Video: Research brings causes of progeria into closer focus:
https://www.youtube.com/watch?v=TjQepjq23EQ

Possible cure for progeria:
Koblan, L. W., Erdos, M. R., Wilson, C., *et al.* (2021). In vivo base editing rescues Hutchinson–Gilford progeria syndrome in mice. *Nature*, 589, 608.

The various times — eons, eras, periods, and epochs — in the development of our planet:
https://geokansas.ku.edu/eons-and-eonothems-periods-and-systems-understanding-how-geologists-talk-about-time

Video: The Geological timescale: https://www.youtube.com/watch?v=r10oh1NHKv4

Video: Earth and the early atmosphere: https://www.youtube.com/watch?v=Gyn754vw8ZQ

Video: Evolution and oxygen: https://www.youtube.com/watch?v=DE4CPmTH3xg

References

Bull, C., and Fee, J. A. (1985). Steady state kinetic studies of superoxide dismutases: properties of the iron containing protein from *Escherichia coli*. *J. Am. Chem. Soc.*, 107, 3295.

Clark, D. S., and Blanch, H. W. (1997). *Biochemical Engineering*, Ed. 2, Marcel Dekker, New York.

DiLoreto, R., and Murphy, C. T. (2015). The cell biology of aging. *Mol. Biol. Cell*, 26, 4525.

Halliwell, B., and Gutteridge, J. M. C. (1999). *Free Radicals in Biology and Medicine*, Ed. 3, Oxford Univ. Press, Oxford.

Halliwell, B., and Gutteridge, J. M. C. (2015). *Free Radicals in Biology and Medicine*, Ed. 5, Oxford Univ. Press, Oxford.

Kizhuveetil, U., Omer, S., Karunagaran, D., and Suraishkumar, G. K. (2020). Improved redox anti-cancer treatment efficacy through reactive species rhythm manipulation. *Sci. Rep.*, 10, 1588.

Menon, K. R., Balan, R., and Suraishkumar, G. K. (2013). Stress induced lipid production in *Chlorella vulgaris*: relationship with specific intracellular reactive species levels. *Biotechnol. Bioeng.*, 110, 1627.

Shin, D. S., DiDonato, M., Barondeau, D. P., Hura, G. L., Hitomi, C., Berglund, J. A., Getzoff, E. D., Cary, S. C., and Tainer, J. A. (2009). Superoxide dismutase from eukaryotic thermophile *Alvinella pompejana*: structures, stability, mechanism, and insights into amyotrophic lateral sclerosis. *J. Mol. Biol.*, 385, 1534.

Chapter 2

Analysis of Reactions that Involve Reactive Species

Valuable insights into the effects of reactive species (RS) are obtained through analyses of their reactions. The essential tools/frameworks available for analysis are thermodynamics and kinetics. Thermodynamic information gives us the possibility of a reaction to occur, and kinetics gives us information on how quickly the reaction occurs. Let us look at them in this chapter.

2.1. Thermodynamics

From the earlier exposures to thermodynamics, let us recall that information on the energies associated with reactions is useful in many ways — from design to operation. The zeroth law of thermodynamics defines temperature as an important parameter, in the context of energy associated with reactions. The first law of thermodynamics essentially states that energy can never be created nor destroyed but can be converted from one form to another. Heat and work must be treated differently because one cannot be converted entirely to another — a consequence of the second law of thermodynamics. The second law of thermodynamics also provides us with a measure to predict the *natural occurrence* of a reaction under a given set of conditions. The measure is entropy.

The useful thermodynamic information for RS is a table of standard reduction potentials. The reactions involving RS, especially ROS, are reduction, oxidation, or reduction–oxidation (redox) reactions. Let us recall that reduction involves a gain in electrons, and oxidation involves a loss of electrons. The species gaining the electron(s) is said to be "reduced," and the species losing the electron(s) is said to be "oxidized." A well-known, nice mnemonic for the same is

LEO, the lion, says GER

LEO stands for Loss of Electrons is Oxidation

GER stands for Gain in Electrons is Reduction

Of course, the addition of oxygen is also called oxidation, and so on. Here, let us focus on electrons. For analysis and use, the reduction and oxidation parts of a redox reaction can be separately written for better conceptual understanding and manipulation. Such reduction or oxidation parts of a redox reaction are called *half-reactions*.

Now, let us consider the conversion of oxygen to water that we saw in Chapter 1. We saw (Figure 1.4) that oxygen is converted to water through the addition of four electrons (along with others such as H).

Let us consider the first reaction in that process:

$$O_2 + e^- \rightarrow O_2^{\bullet-}$$

This is a reduction reaction because it involves the addition of an electron. If this reaction is part of an electrochemical cell with a standard electrode, we can measure the voltage and, thus, the energy associated with this reaction.

Let us recall that the standard electrode is a platinum electrode dipped in a 1 M solution of H^+ ions and exposed to pure hydrogen (H_2) gas at 1 atm pressure and 298 K. The following reversible reaction takes place:

$$0.5\ H_2 \rightleftharpoons H^+ + e^-$$

The potential of the standard electrode is assigned the value of zero to serve as a reference. The standard reduction potential of a half-reaction, E^0, is the voltage measured when the half-reaction under consideration is paired with a standard hydrogen electrode.

For example, if the above reduction reaction, $O_2 + e^- \rightarrow O_2^{\bullet-}$, represented as $O_2/O_2^{\bullet-}$ (a half-reaction in this context), is coupled with the half-reaction of a standard hydrogen electrode in electrochemical cell at physiological pH (pH 7), the electrical potential difference associated with the reaction is −0.33 V. This assumes that all chemical entities of relevance, except hydrogen ions, are present at concentrations of 1 M, each. Since we are interested in biological systems, the pH (calculated as the negative base 10 logarithm of the hydrogen ion concentration) of interest is the physiological pH, 7. The standard reduction potential at pH 7 is represented as $E^{0\prime}$ and is useful thermodynamic information. Tables of standard reduction potentials at pH 7 are available in textbooks and the literature.

Table 2.1 is a sample table of reduction potentials. The half-reactions are also represented as (species with a smaller number of electrons)/(species with a larger number of electrons). For example, we saw above that

$$O_2 + e^- \rightarrow O_2^{\bullet-} \text{ is represented as } O_2/O_2^{\bullet-}$$

A half-reaction with a −ve $E^{0\prime}$ value will have a *tendency* to release electrons. However, irrespective of +ve or −ve values, a half-reaction has the ability to reduce all systems with an $E^{0\prime}$ value higher than it.

A half-reaction with a +ve $E^{0\prime}$ value will have a *tendency* to accept electrons. However, irrespective of +ve or −ve values, a half-reaction has the ability to oxidize all systems with an $E^{0\prime}$ value lower than it.

Note that $E^{0\prime}$ refers to the standard value, that is, at the following conditions:

$$T = 298 \text{ K}$$
$$P = 1 \text{ atm}$$

Concentrations = 1 M for all species involved except hydrogen.

The temperature and concentrations of interest may often be different from the above standard values. The effective reduction potential at actual conditions, E', can be found by using the Nernst equation that you would have seen in your earlier classes:

$$E' = E^{0\prime} + \frac{RT}{nF}\ln\frac{[species\ with\ lesser\ number\ of\ electrons]}{[species\ with\ higher\ number\ of\ electrons]} \qquad \text{Eq. 1}$$

where R is the gas constant, n is the number of moles of electrons transferred, F is the Faraday's constant, 96,485 C (mole of electrons)$^{-1}$. The square brackets represent the concentrations of the relevant species they hold.

The electron accepted by oxygen in the reaction, $O_2 + e^- \rightarrow O_2^{\bullet-}$, can be transferred to other suitable species. The transfer could be through a single step or multiple steps. An example of electron transfer in a single step, as we have seen in Chapter 1, is the transfer to H_2O_2:

$$O_2^{\bullet-} + e^- + 2H^+ \rightarrow H_2O_2$$

An example of electron transfer through multiple steps, which we have already seen in Chapter 1, is the transfer to H_2O, the last step being:

$$^{\bullet}OH + e^- + H^+ \rightarrow H_2O$$

Suppose we have the value of the reduction potential of the above reactions. In that case, the Gibbs free energy or the energy generated/consumed in the process of electron transfer can be calculated through the relationship that you would already know:

$$\Delta G^{0'} = -nF\,\Delta E^{0'} \qquad\qquad \text{Eq. 2}$$

Equation 2 can be used if the concentration of all species involved at physiological pH is 1 M.

$$\Delta G' = -nF\,\Delta E' \qquad\qquad \text{Eq. 3}$$

Equation 3 can be used if, at physiological pH, the concentration of each involved species is **not** 1 M.

Let us illustrate. From Table 2.1 of reduction potentials, the standard reduction potential at pH 7, $E^{0'}$, of

$$O_2/O_2^{\bullet-} \qquad \text{is } -0.33 \text{ V}$$

$$O_2^{\bullet-}/H_2O_2 \qquad \text{is } 0.94 \text{ V}$$

Note that the number of electrons transferred, n, needs to be known for the calculation of ΔG. Recall that ΔG is an extensive property, that is, it depends on the amount of substance present. But, note that the reduction

Table 2.1.   Standard reduction of potentials of some biological half-reactions at pH 7

Half-reaction representation	Half-cell representation	Standard reduction potential at pH 7, E_o' (V)
$H_2O \rightarrow H_2O^+ + e^-_{aq}$ (*hydrated electron*)	H_2O/e^-_{aq}	−2.840
$CO_2 + e^- \rightarrow CO_2^{\bullet-}$	$CO_2/CO_2^{\bullet-}$	−1.800
$O_2 + H^+ + e^- \rightarrow HO_2^{\bullet}$	$O_2/HO_2^{\bullet-}$	−0.460
menadione $+ e^- \rightarrow$ *menadione*$^{\bullet-}$	*menadione/menadione*$^{\bullet-}$	−0.450
ferredoxin$(Fe^{3+}) + e^- \rightarrow$ *ferrdoxin* (Fe^{2+})	*ferredoxin*(Fe^{3+})/*ferrdoxin*(Fe^{2+})	−0.432
$2H^+ + 2e^- \rightarrow H_2$ (*note: pH = 7*)	$2H^+/H_2$	−0.414
α-ketoglutarate $+ CO_2 + 2H^+ + 2e^- \rightarrow$ *isocitrate*	*α-ketoglutarate/isocitrate*	−0.380
acetoacetate $+ 2H^+ + 2e^- \rightarrow$ *β-hydroxybutyrate*	*acetoacetate/β-hydroxybutyrate*	−0.346
$O_2 + e^- \rightarrow O_2^{\bullet-}$	$O_2/O_2^{\bullet-}$	−0.330
$NADP^+ + H^+ + 2e^- \rightarrow NADPH$	$NADP^+/NADPH$	−0.324
$NAD^+ + H^+ + 2e^- \rightarrow NADH$	$NAD^+/NADH$	−0.320
lipoic acid $+ 2H^+ + 2e^- \rightarrow$ *dihydrolipoic acid*	*lipoic acid/dihydrolipoic acid*	−0.290
$S + 2H^+ + 2e^- \rightarrow H_2S$	S/H_2S	−0.243
$GSSG + 2H^+ + 2e^- \rightarrow GSH$	$GSSG/GSH$	−0.230
acetaldehyde $+ 2H^+ + 2e^- \rightarrow$ *ethanol*	*acetaldehyde/ethanol*	−0.197
ferritin$(Fe^{3+}) + e^- \rightarrow$ *ferritin*(Fe^{2+})	*ferritin*(Fe^{3+})/*ferritin*(Fe^{2+})	−0.190
pyruvate $+ 2H^+ + 2e^- \rightarrow$ *lactate*$^-$	*pyruvate/lactate*$^-$	−0.185
dihydroascorbate $+ e^- \rightarrow$ *ascorbate*$^{\bullet-}$	*dihydroascorbate/ascorbate*$^{\bullet-}$	−0.170

(Continued)

Table 2.1. *(Continued)*

Half-reaction representation	Half-cell representation	Standard reduction potential at pH 7, E_o' (V)
$oxaloacetate^{2-} + 2H^+ + 2e^- \rightarrow malate^{2-}$	$oxaloacetate^{2-}/malate^{2-}$	-0.166
$fumarate^{2-} + 2H^+ + 2e^- \rightarrow succinate^{2-}$	$fumarate^{2-}/succinate^{2-}$	0.031
$ubiquinone + 2H^+ + 2e^- \rightarrow ubiquinol + H_2$	$ubiquinone/ubiquinol$	0.045
$cytochrome\ b\ (Fe^{3+}) + e^- \rightarrow cytochrome\ b\ (Fe^{2+})$	$cytochrome\ b\ (Fe^{3+})/cytochrome\ b\ (Fe^{2+})$	0.077
$(Fe^{3+} \sim ADP) + e^- \rightarrow (Fe^{2+} \sim ADP)$	$(Fe^{3+} \sim ADP)/(Fe^{2+} \sim ADP)$	~0.100
$(Fe^{3+} \sim citrate) + e^- \rightarrow (Fe^{2+} \sim citrate)$	$(Fe^{3+} \sim citrate)/(Fe^{2+} \sim citrate)$	~0.100
$(Fe^{3+} \sim EDTA) + e^- \rightarrow (Fe^{2+} \sim EDTA)$	$(Fe^{3+} \sim EDTA)/(Fe^{2+} \sim EDTA)$	0.120
$cytochrome\ c_1(Fe^{3+}) + e^- \rightarrow cytochrome\ c_1(Fe^{2+})$	$cytochrome\ c_1(Fe^{3+})/cytochrome\ c_1(Fe^{2+})$	0.220
$cytochrome\ c(Fe^{3+}) + e^- \rightarrow cytochrome\ c(Fe^{2+})$	$cytochrome\ c(Fe^{3+})/cytochrome\ c(Fe^{2+})$	0.254
$ascorbate^{\bullet-} + H^+ + e^- \rightarrow ascorbate^-$	$ascorbate^{\bullet-}/ascorbate^-$	0.280
$cytochrome\ a(Fe^{3+}) + e^- \rightarrow cytochrome\ a(Fe^{2+})$	$cytochrome\ a(Fe^{3+})/cytochrome\ a(Fe^{2+})$	0.290
$O_2 + 2H^+ + 2e^- \rightarrow H_2O_2$	O_2/H_2O_2	0.295
$NAD^{\bullet} + H^+ + e^- \rightarrow NADH$	$NAD^{\bullet}/NADH$	0.300
$H_2O_2 + H^+ + e^- \rightarrow {}^{\bullet}OH + H_2O$	$H_2O_2/{}^{\bullet}OH$	0.320
$Fe(CN)_6^{3-} + e^- \rightarrow Fe(CN)_6^{4-}$	$Fe(CN)_6^{3-}/Fe(CN)_6^{4-}$	0.360

Reaction	Couple	Value
$cytochrome\ f(Fe^{3+}) + e^- \rightarrow cytochrome\ f(Fe^{2+})$	$cytochrome\ f(Fe^{3+})/cytochrome\ f(Fe^{2+})$	0.365
$NO_3^- + 2H^+ + 2e^- \rightarrow NO_2^- + H_2O$	NO_3^-/NO_2^-	0.421
$\alpha\text{-}tocopherol^{\bullet} + H^+ \rightarrow \alpha\text{-}tocopherol$	$\alpha T^{\bullet}/\alpha TH$	0.500
$H\text{-}urate^{\bullet-} + H^+ \rightarrow urate^-$	$HU^{\bullet-}/UH_2^-$	0.590
$RO_2^{\bullet} + H^+ \rightarrow ROOH$	$RO_2^{\bullet}/ROOH$	$0.770 - 1.440$ *for dif R*
$Fe^{3+} + e^- \rightarrow Fe^{2+}$	Fe^{3+}/Fe^{2+}	0.771
$O_2 + 2H^+ + 2e^- \rightarrow 2H_2O$	O_2/H_2O	0.816
$cysteine^{\bullet} + e^- \rightarrow cysteine$	$RS^{\bullet}/RS^-$	0.920
$O_2^{\bullet-} + 2H^+ + e^- \rightarrow H_2O_2$	$O_2^{\bullet-}/H_2O_2$	0.940
$HO_2^{\bullet} + H^+ + e^- \rightarrow H_2O_2$	$HO_2^{\bullet}/H_2O_2$	1.060
$RO^{\bullet} + H^+ + e^- \rightarrow ROH(aliphatic\ alkoxyl)$	$RO^{\bullet}/ROH$	1.600 (*R dep*)
$^{\bullet}OH + H^+ \rightarrow H_2O$	$^{\bullet}OH/H_2O$	2.310

Data compiled from various sources: Halliwell and Gutteridge [2015], Buettner [1993], Loach [1976]. The data from different literature sources slightly varied.

potential, $E^{0\prime}$, is an intensive property, that is, it does not depend on the amount of substance present; $E^{0\prime}$ is the potential difference created between two electrodes. Also, note that ΔG is a state function — its value depends on the "state," or the set of T, P, and specific volume combination (recall "equation of state"), whereas $E^{0\prime}$ is not a state function.

Let us first consider the process of H_2O_2 formation from oxygen. It can be represented by the following chemical equation:

$$O_2 + 2e^- + 2H^+ \rightarrow H_2O_2$$

The above can be obtained by adding the following half-reactions, represented by Eqs. A and B, which are present in the table of standard reduction potentials.

$$O_2 + e^- \rightarrow O_2^{\bullet-} \qquad\qquad \text{Eq. A}$$
$$O_2^{\bullet-} + e^- + 2H^+ \rightarrow H_2O_2 \qquad\qquad \text{Eq. B}$$

Thus, we could use the energy information available in that table.

Since the reduction potential is not a state function, we cannot merely add the reduction potentials of Eqs. A and B to get the reduction potential for the needed process. However, ΔG is a state function, and thus, the ΔG values can be added, in this case, to get the ΔG value for the needed process. Thus,

$$O_2 + e^- \rightarrow O_2^{\bullet-}$$
$$E^{0\prime} = -0.33\,V, \quad \Delta G^{0\prime} = -nFE^{0\prime} = -1 \times 96{,}485 \times (-0.33) = 31{,}840.05\ J$$
$$O_2^{\bullet-} + e^- + 2H^+ \rightarrow H_2O_2$$
$$E^{0\prime} = 0.94\,V, \quad \Delta G^{0\prime} = -nFE^{0\prime} = -1 \times 96{,}485 \times (0.94) = -90{,}695.9\ J$$

Thus, for the needed process,

$$O_2 + 2e^- + 2H^+ \rightarrow H_2O_2$$

The standard free energy available is (Since the reactions A and B were added together to get the needed process, the standard free energy values of reactions A and B also need to be added together to get the needed free energy change for the process)

$$\Delta G^{0\prime} = 31{,}840.05 + (-90{,}695.9) = -58{,}855.85\ J = -5.886 \times 10^4\ J$$

Let us recall that (Coulomb − Volt) = Joule. In electrical terms, 1 J is the amount of energy needed to transport a charge of 1 C through a potential difference of 1 V.

Also, the value of the Gibbs free energy of the process is negative. The negative value tells us that the process is spontaneous.

One needs to be careful while using reduction potentials to get free energy information. While dealing with redox reactions, *the number of electrons in half-reactions may need balancing*. Also, appropriate algebraic manipulations may be required to get the desired process from the half-reactions available in the reduction potentials table. More importantly, one should realize that the Gibbs free energy is a state function, whereas the reduction potential is not a state function. Thus, only Gibbs free energies can be algebraically manipulated, not the reduction potentials.

Next, let us consider the interaction between cytochrome c and $FADH_2$. Under the relevant conditions in the cell, say the concentration of (Fe^{3+})-cytochrome c is 10^{-6} M, (Fe^{2+})-cytochrome c is 3×10^{-6} M, $FADH_2$ is 10^{-4} M, and FAD is 2×10^{-4} M. Find the available free energy when electrons are transferred from $FADH_2$ to cytochrome c.

From the table of standard reduction potentials,

$$FAD + 2H^+ + 2e^- \rightarrow FADH_2 \qquad Eq.\ C \qquad E^{0'} = -0.18\ V$$

$$(Fe^{3+})cytochrome\ c + e^- \rightarrow (Fe^{2+})cytochrome\ c \qquad Eq.\ D \qquad E^{0'} = 0.26\ V$$

Let us first find the standard available free energy and then the actual available free energy. To find the standard available free energy, we can follow the same procedure as in the previous example with oxygen and H_2O_2. However, Eqs. C and D together represent a redox reaction. Thus, we need to ensure that the number of electrons is balanced. As written, the reaction with FAD has two electrons, whereas the reaction with cytochrome c has one electron. Since we are considering the process of electron transfer from $FADH_2$ and cytochrome c, the electrons need to be balanced.

We can balance the number of electrons in this case by multiplying the cytochrome c reaction by a factor of 2 to result in:

$$2(Fe^{3+})cytochrome\ c + 2e^- \rightarrow 2(Fe^{2+})cytochrome\ c \qquad Eq.\ E \qquad E^{0'} = 0.26\ V$$

Note that the standard reduction potential, $E^{0\prime}$, was left unchanged — it was not multiplied by the factor 2. That is because $E^{0\prime}$ is an intrinsic quantity; recall that it is a potential difference between the two electrodes, which does not change with the number of electrons transferred.

Now, the needed process of electron transfer from $FADH_2$ to cytochrome b can be obtained through (Eqs. E–C). The electrons will cancel out with that algebraic operation. Thus, the standard available free energy is

$$-nFE^{0\prime}\left(eq.E\right)-\left(-nFE^{0\prime}\left(eq.C\right)\right)=-2F\left[E^{0\prime}\left(eq.E\right)-E^{0\prime}\left(eq.C\right)\right]$$

$$=-2F\Delta E^{0\prime}$$

$$=-2F(0.26-(-0.18))=-2\times 96{,}485\times 0.44=-8.491\times 10^{4}\,J$$

Now, the actual available free energy is needed (the species concentrations are not all 1 M at pH 7). Let us recall Eq. 1,

$$E' = E^{0\prime} + \frac{RT}{nF}\ln\frac{\left[species\ with\ lesser\ number\ of\ electrons\right]}{\left[species\ with\ higher\ number\ of\ electrons\right]}$$

If we assume a mammalian cell (T = 37°C = 310 K), for the FAD half-reaction,

$$E' = E^{0\prime} + \frac{RT}{nF}\ln\frac{\left[FAD\right]}{\left[FADH_2\right]}=-0.18+\frac{8.31\times 310}{2\times 96{,}485}\ln\frac{\left[2\times 10^{-4}\right]}{\left[10^{-4}\right]}=-0.171V$$

And, for the cytochrome c half-reaction

$$E' = E^{0\prime} + \frac{RT}{nF}\ln\frac{\left[\left(Fe^{3+}\right)cytochrome\ c\right]}{\left[\left(Fe^{2+}\right)cytochrome\ c\right]}$$

$$=0.26+\frac{8.31\times 310}{2\times 96{,}485}\ln\frac{\left[10^{-6}\right]}{\left[3\times 10^{-6}\right]}=0.2453V$$

Therefore, on the same lines as the standard case, the available free energy in this case is

$$-nFE'(eq.E) - (-nFE'(eq.C)) = -2F[E'(eq.E) - E'(eq.C)] = -2F\Delta E'$$

$$= -2F(0.2453 - (-0.171)) = -2 \times 96{,}485 \times 0.4163 = -8.033 \times 10^4 \, J$$

2.2. Kinetics

The other important tool available for the analysis of RS reactions is kinetics. Let us review the following fundamental concepts in kinetics.

Rate refers to the speed of a reaction/process. It has an instantaneous connotation, although it could be measured in different ways. The rate is usually normalized. The normalizing factor could be the relevant volume, area, mass, and so on, of the containing substance/matrix.

$$rate = \frac{1}{time} \frac{change\ in\ amount\ of\ the\ species}{normalizing\ factor}$$

Note that the reaction rate is not always equal to $\frac{dC}{dt}$, which is the accumulation rate. To understand better, recall that the fundamental mass balance on a species is:

$$input\ rate + generation\ rate - output\ rate - consumption\ rate =$$

$$accumulation\ rate,\ \frac{dm}{dt} = \frac{d(VC)}{dt}$$

m = mass of the species being balanced
C = volumetric concentration of the species being balanced
V = volume of the system

The strategy used to measure reaction rate is to measure concentrations of a relevant species, a reactant or a product in the reaction at various times in a well-mixed vessel operating in a batch mode. In the mass balance of a species over such a system, there is no input rate, no output rate. Thus, the net production of a product or the net consumption rate of a reactant/substrate equals the accumulation rate of that species. The reaction rate equals the accumulation rate only for a batch process.

The **rate law** refers to the mathematical expression relating the reaction rate to the species concentrations (note that concentrations are usually easy to measure).

$rate = k\,(concentration\ of\ species\ 1)^{n1}\ (concentration\ of\ species\ 2)^{n2}\ldots$

k = **rate constant**

$n_1,\ldots$ = order with respect to that species; $n = \sum_{ni}$ is the overall order

Now, let us look at some RS reactions in the context of initiation, propagation, and termination reactions, which are relevant to FR reactions. It also directly provides a basis to analyze free radical reactions. Recall that only some RS are free radicals (FR). There are many non-radical RS, as we saw earlier.

2.2.1. *Some initiation reactions*

Fenton reaction: Let us begin with a well-known reaction to initiate hydroxyl radicals, namely the Fenton reaction. The reaction between H_2O_2 and metal ions such as Fe^{2+} or Cu^+ is called the Fenton reaction. Those reactions are:

$$H_2O_2 + Fe^{2+} \rightarrow Fe^{3+} + OH^- + {}^{\bullet}OH \qquad k = 76\ M^{-1}s^{-1}$$

$$H_2O_2 + Cu^+ \rightarrow Cu^{2+} + OH^- + {}^{\bullet}OH \qquad k = 4.7 \times 10^3\ M^{-1}s^{-1}$$

Some variants of the above reactions are also called Fenton's reaction. For example, in the cell, Fe^{2+} also gets chelated to citrate (Fe^{2+} – citrate) or ATP (Fe^{2+} – ATP) and other molecules. The reaction between Fe^{2+} – citrate (or Fe^{2+} – ATP) and H_2O_2 is also called Fenton's reaction. The rate constant for the chelate with H_2O_2 is much higher, that is, $1.6 \times 10^4\ M^{-1}\ s^{-1}$ at 37°C for Fe^{2+} – ATP, and an estimated $1.2 \times 10^4\ M^{-1}\ s^{-1}$ for Fe^{2+} – citrate.

In normal physiology, typical intracellular concentrations of H_2O_2 and Fe^{2+} – citrate are low at 10^{-8} M and 10^{-6} M, respectively.

Example: Estimate the number of hydroxyl molecules produced per sec per cell due to a Fenton reaction under the above typical conditions.

Solution:

$$rate = k[H_2O_2]\,[Fe^{2+} - citrate]$$
$$= 1.2 \times 10^4(10^{-8})(10^{-6}) = 1.2 \times 10^{-10} \; Ms^{-1} \; or \; mole \; L^{-1}s^{-1}$$

Although the rate may seem low, let us calculate the number of hydroxyl radicals formed per second through the Avagadro number.

$$1.2 \times 10^{-10} \; mole \; L^{-1} \; s^{-1} = 1.2 \times 10^{-10} \times 6.023 \times 10^{23} \; molecules \; L^{-1} \; s^{-1}$$
$$= 7.2 \times 10^{13} \; molecules \; L^{-1} \; s^{-1}$$

A typical mammalian cell diameter is of the order of 10 μm. Thus, its volume is of the order of 10^{-15} m^3 or 10^{-12} L. Thus

$$7.2 \times 10^{13} \times 10^{-12} = 72 \; molecules \; s^{-1}$$

Seventy-two hydroxyl molecules are produced in a cell every second, which is substantial enough to cause damage.

Iron chelates can react with HOCl, a water disinfectant, to produce hydroxyl radicals — of relevance in water treatment.

$$Fe^{2+} - chelate + HOCl \rightarrow Fe^{3+} - chelate + Cl^- + {}^{\bullet}OH$$
$$k = 1.3 \times 10^4 \; M^{-1}s^{-1} \; at \; 25 \; °C$$

Photolysis: Photolysis refers to the breakage of bonds using light. Breakage using light (UV or other wavelengths) results in homolytic (equal) fission of a covalent bond, with each species taking an unpaired electron, which results in FR. For example, UV-induced homolytic fission of the O–O bond in H_2O_2 results in hydroxyl radicals.

$$HO - OH \xrightarrow{\;uv\;} 2\,{}^{\bullet}OH$$

H_2O_2 is produced as a part of the normal metabolic processes in the cell — many enzymes such as xanthine oxidase, urate, SOD, and others generate H_2O_2. Some other RS reactions also generate H_2O_2. If the skin is exposed too long to sunlight, the UV rays can penetrate the skin into the skin cells, generate hydroxyl radicals, and cause cell damage, especially DNA damage. It results in skin cancer — basal cell carcinoma or malignant melanoma.

In the troposphere, hydroxyl and nitric oxide RS are generated by the photolysis of nitrous acid.

$$HNO_2 \xrightarrow{\ light\ } {}^{\bullet}OH + NO^{\bullet}$$

The RS formed can readily oxidize other organic compounds in the atmosphere and thus contribute to atmospheric pollution.

Radiolysis: Radiolysis refers to the use of high-energy radiation to split water. For example,

$$H_2O \xrightarrow{\ \gamma-rays\ } H^{\bullet} + {}^{\bullet}OH$$

This happens during radiolysis, one of the techniques used to measure the kinetics of FR reactions, as we will see later.

Ultrasound: Ultrasound or high-frequency vibrations cause the formation, growth, and collapse of gas bubbles in a liquid (see Additional Information). This leads to microscopic pockets of high temperature (hundreds of degrees C) and pressure (hundreds of atmospheres) for very short time periods that can erode even metal; this process is called cavitation. Under such conditions, homolytic fission of water can occur through a different mechanism from the above radiation-induced homolysis to yield the same hydrogen and hydroxyl FR. If this happens in a biological system, it may lead to cytotoxicity and is undesirable. For example, the following procedures need to be carefully evaluated:

Phacoemulsification: The removal of cataracts using an ultrasonic probe to shatter the eye lens. The broken-up eye lens is removed by aspiration (see Additional Information). During phacoemulsification, it is highly likely that the hydroxyl radicals formed by the homolysis of water cause cell damage. Thus, a solution containing hyaluronic acid, a good antioxidant against hydroxyl species, is used during the process.

Shock wave lithotripsy: Breaking kidney stones through rapidly pulsed sonic waves (see Additional Information) — this is known to cause cytotoxicity in animal studies.

Liposome preparation: The lipids are sonicated during the preparation of liposomes — one needs to be careful about the possible effects of generated RS.

2.2.2. *Some propagation reactions*

Now, let us look at some notable reactions that propagate FR — that is, FR are present on both sides of the reaction as reactants and products. In the initiation reactions, FR were present only as reaction products.

Hypochlorous acid has significance in water treatment, as we have briefly discussed earlier. Further, HOCl reacts with superoxide to produce hydroxyl radicals.

$$HOCl + O_2^{\bullet -} \rightarrow O_2 + Cl^- + {}^\bullet OH$$
$$k = 7.5 \times 10^6 \, M^{-1}s^{-1} \text{ at } 25 \, ^\circ C$$

Some of the widely found reactions of the hydroxyl radical are propagation reactions: hydrogen abstraction, addition, and electron transfer reactions.

Hydrogen abstraction: When hydroxyl radical reacts with ethanol, it abstracts a hydrogen radical and forms water.

$$C_2H_5OH + {}^\bullet OH \rightarrow (CH_3HCOH)^\bullet + H_2O$$

When the above carbon radical reacts with oxygen, a peroxyl radical results:

$$CH_3HC^\bullet OH + O_2 \rightarrow (O_2CH_3HCOH)^\bullet$$

Addition: A species with an unpaired electron (radical) may add to another molecule. Since electrons are conserved, the additive product (adduct) would still have the unpaired electron. For example,

$$\bullet OH + \quad \text{guanine} \quad \longrightarrow \quad \text{G8OH}$$

guanine   G8OH

Guanine is a purine. This also happens with thymine, a pyrimidine. Thymine can further react with oxygen to give thymine peroxyl radicals.

Electron transfer: The hydroxyl radical participates in electron transfers, for example, with halide ions:

$$Cl^- + {}^\bullet OH \rightarrow Cl^\bullet + OH^-$$

The chloride radical can further react with chloride ion to give:

$$Cl^- + Cl^\bullet \rightarrow Cl_2^{\bullet -}$$

Another electron transfer reaction that is biologically relevant is with nitrite ion:

$$NO_2^- + {}^\bullet OH \rightarrow NO_2^\bullet + OH^-$$

pH buffer system in blood and cell culture:
An important propagation reaction in the body involves the carbonate (and bicarbonate) ion. As may be known, the bicarbonate (HCO_3^-) level in the blood plasma is 25 mM, and it serves to buffer the blood against pH changes. Also, the mammalian cell culture medium uses the same buffer system. Let us first look at how the buffering takes place.

The carbon dioxide from the atmospheric air enters the human body through the respiration system. It is converted to carbonic acid under the action of the enzyme, carbonic anhydrase, and carbonic acid dissociates further in the blood.

$$CO_2 + H_2O \xleftrightarrow{\;carbonic\ anhydrase\;} H_2CO_3 \leftrightarrow HCO_3^- + H^+$$

The buffer needs to resist pH changes on addition of acids or bases to itself. Acid addition is a common occurrence due to cell metabolism, and so on. In this case, when an acid is added, the acid reacts with the bicarbonate ion (right-most reaction) and produces more carbonic acid (drives the reaction to the left) so that the H^+ level, hence the pH, remains constant. In humans, this is further coupled with the respiration rate, which adjusts the amount of CO_2 (leftmost reaction) to take care of the acidification of the blood. More CO_2 is produced when an acid (or H^+) is added, and the increased respiration rate removes the excess CO_2. The addition of a base works the reaction in the other direction; the base reacts with H^+ to push the reactions to the right.

Now, let us see some propagation reactions that involve the radicals or ions relevant to the above system.

$$HCO_3^- + {}^\bullet OH \rightarrow H_2O + CO_3^{\bullet -} \qquad\qquad k \approx 10^7 \; M^{-1}s^{-1}$$
$$CO_3^{2-} + {}^\bullet OH \rightarrow OH^- + CO_3^{\bullet -} \qquad\qquad k \approx 2 \times 10^8 \; M^{-1}s^{-1}$$

The carbonate radical seen above is also formed through other reactions in the body/cell. It is a potent one-electron oxidizing agent ($E^{0\prime} = 1.78$ V) and oxidizes several biomolecules, including hyaluronic acid. It can abstract a hydrogen radical from cysteine, tyrosine, or NADH.

$$cys - SH + CO_3^{\bullet -} \rightarrow cys - S^\bullet + HCO_3^- \qquad k = 4.6 \times 10^7 \; M^{-1}s^{-1}$$
$$Tyr - OH + CO_3^{\bullet -} \rightarrow Tyr - O^\bullet + HCO_3^- \qquad k = 4.5 \times 10^7 \; M^{-1}s^{-1}$$
$$NADH + CO_3^{\bullet -} \rightarrow NAD^\bullet + HCO_3^- \qquad k = 1.4 \times 10^9 \; M^{-1}s^{-1}$$

The NAD radical can further react with oxygen to yield superoxide.

$$NAD^\bullet + O_2 \rightarrow NAD^+ + O_2^{\bullet -} \qquad\qquad k = 1.9 \times 10^9 \; M^{-1}s^{-1}$$

Thiols (R–SH) can act as antioxidants. Glutathione (GSH), a tripeptide, Glu–Cys–Gly, is a cellular antioxidant, and it has been hypothesized that hydrogen sulfide, H–SH, is an antioxidant in the brain. On accepting the unpaired electron from another molecule, they become FR. However, these thiol FR are relatively much more stable, and thus, they act as antioxidants. For example,

$$R - SH + {}^\bullet OH \rightarrow RS^\bullet + H_2O$$

The $RS^{\bullet}$ is called the thiyl radical.

$$R - SH + R'O_2^{\bullet} \rightarrow RS^{\bullet} + R'OOH$$

When an RS abstracts a hydrogen atom from a hydrocarbon side chain of a fatty acid residue, a $\equiv C^{\bullet}$ results. That free radical can interact with a $R - SH$ as follows:

$$R - SH + \equiv C^{\bullet} \rightarrow RS^{\bullet} + \equiv CH$$

Thiyl radicals are formed when some metal ions react with $R - SH$:

$$R - SH + Fe^{3+} \rightarrow RS^{\bullet} + Fe^{2+} + H^+$$
$$R - SH + Cu^{2+} \rightarrow RS^{\bullet} + Cu^{2+} + H^+$$

Thiyl radicals are also formed by homolytic fission of disulfide bonds in proteins:

$$cysteine - S - S - cysteine \xrightarrow{\ homolytic\ fission\ } cysteine - S^{\bullet} + S^{\bullet} - cysteine$$

This may result in the food industry when flour is milled. The cysteine thiyl radical ($Cys - S^{\bullet}$) can cause lipid peroxidation by the abstraction of hydrogen from linoleic, linolenic, and arachidonic acids with rate constants of 10^6 to 10^7 $M^{-1}s^{-1}$.

Thiyl radical reacts with ascorbate as follows:

$$GS^{\bullet} + AH^- \rightarrow GSH + A^{\bullet -} \qquad\qquad k \approx 5 \times 10^8\ M^{-1}s^{-1}$$

It also reacts with NADH and NAD(P)H to yield NAD radical:

$$GS^{\bullet} + NADH \rightarrow GS^- + NAD^{\bullet} + H^+ \qquad\qquad k \approx 2.3 \times 10^8\ M^{-1}s^{-1}$$

Before we consider some termination reactions, let us look at a special type of reaction called the dismutation reaction. When the same species is both oxidized and reduced, the species is said to be dismutated. For example, the dismutation of superoxide is represented as

$$O_2^{\bullet-} + O_2^{\bullet-} + 2H^+ \rightarrow O_2 + H_2O_2$$

The superoxide dismutase enzyme, which we can represent as E – Cu, can mediate this dismutation. The explicit oxidation and reduction parts of the above dismutation of superoxide are as follows:

$$\text{Oxidation: } E - Cu^{2+} + O_2^{\bullet-} \rightarrow E - Cu^+ + O_2$$

$$\text{Reduction: } E - Cu^+ + O_2^{\bullet-} + 2H^+ \rightarrow E - Cu^{2+} + H_2O_2$$

The dismutation of superoxide can also be mediated by Mn, which like Cu, is a transition metal needed in the cell for the activity of several enzymes such as SOD.

2.2.3. *Some termination reactions*

When two hydroxyl radicals meet, they form H_2O_2:

$$^{\bullet}OH + {}^{\bullet}OH \rightarrow H_2O_2 \qquad\qquad k = 5 \times 10^9 \; M^{-1}s^{-1}$$

If the rate constants are large, one radical can be used to scavenge another, more dangerous radical. For example, $NO^{\bullet}$ can be used to scavenge hydroxyl and peroxyl radicals:

$$NO^{\bullet} + {}^{\bullet}OH \rightarrow HNO_2 \qquad\qquad k > 10^{10} \; M^{-1}s^{-1}$$

$$NO^{\bullet} + RO_2^{\bullet} \rightarrow ROONO \qquad\qquad k > 10^9 \; M^{-1}s^{-1}$$

Nitric oxide has multiple physiological roles in the vascular and nervous systems. For example, $NO^{\bullet}$ synthesized by the vascular endothelial cells that line the interior of the blood vessels, reaches the underlying smooth muscle by diffusion, binds to guanylate cyclase and activates it. This results in more cGMP production, which lowers intracellular Ca^{2+}, relaxes the muscle, dilates the vessel, and thus lowers the blood pressure.

Thiyl radical disappearance by dimerization,

$$GS^{\bullet} + GS^{\bullet} \rightarrow GSSG \qquad\qquad k = 1.5 \times 10^9 \; M^{-1}s^{-1}$$

which is usually assumed because the reaction rate constant is high, is not likely. This is because the cellular concentration of $GS^{\bullet}$ is low enough to preclude interaction between them. Nevertheless, GSH is present in mM levels in the cell, and the pKa of the thiol group of GSH is 9.2. So it is about 1–2% ionized at pH 7.4, the intracellular pH of many mammalian cells.

$$GSH \rightleftharpoons GS^- + H^+$$

Ionized glutathione can react rapidly with $GS^{\bullet}$:

$$GS^{\bullet} + GS^- \rightarrow GSSG^{\bullet-} \qquad\qquad k = 8 \times 10^8\ M^{-1}s^{-1}$$

Although $GS^{\bullet}$ is a good oxidizing agent (reduction potential ~0.9 V), $GSSG^{\bullet-}$ is a good reducing agent (reduction potential ~ −1.5 V). It can reduce oxygen to superoxide

$$GSSG^{\bullet-} + O_2 \rightarrow GSSG + O_2^{\bullet-} \qquad\qquad k = 5 \times 10^8\ M^{-1}s^{-1}$$

2.2.4. *Other RS that are not free radicals*

Other RS, such as peroxynitrite, hydrogen peroxide, HOCl, singlet oxygen, and ozone, are not FR. Further, some propagation reactions involving FRs could lead to the formation of FRs of interest, and hence, they need to be considered as "initiation" ones. These need to be taken into account if the analysis needs it.

2.2.5. *How are rate constants measured in RS reactions?*

Two methods that are commonly employed to measure rate constants of reactions that involve RS are:

- Pulse radiolysis
- Stopped flow

2.2.5.1. *Pulse radiolysis*

Pulse radiolysis is used to determine the kinetics of very fast reactions. It can be used to study the kinetics of any species (say, A), especially biological molecules, with

- hydroxyl radical, or
- superoxide radical

First, a pulse (10^{-6} to 10^{-8} s) of high energy electrons from a linear accelerator is provided to a reaction cell that contains the species/compound of interest, A, in an aqueous solution (sometimes in organic solvents). The A-derived radicals are monitored over short times (10^{-9} to 10 s) through absorbance, fluorescence, or EPR spectroscopy. The other concentrations needed to calculate the second-order reaction-rate constants are measured by competition methods, as discussed later.

When the reaction mixture is pulsed with high-energy electrons, the following takes place:

In the first 10^{-16} s: absorption of energy by water molecules to result in water ions and activated water species:

$$2H_2O \rightarrow H_2O^+ + e^- + H_2O^*$$

In 10^{-14} to 10^{-13} s: homolytic fission of the activated water species and the reaction of the water ions with water molecules.

$$H_2O^* \rightarrow H^\bullet + {}^\bullet OH$$

$$H_2O^+ + H_2O \rightarrow H_2O^+ + {}^\bullet OH$$

In 10^{-12} to 10^{-11} s: the electrons generated by the first in the above sequence of reactions get surrounded by clusters of water molecules. These are called hydrated electrons, e^-_{aq}.

The three radicals that result from the high energy pulse, $H^\bullet$, ${}^\bullet OH$, and e^-_{aq} are first formed in clusters, micro-regions of high radical concentrations, called spurs. Radical recombinations within these spurs to produce H_2O_2 and H_2 are complete in 10^{-8} s.

The hydroxyl or superoxide radicals that react with A are chosen as follows:

- If hydroxyl is needed, then before the pulse exposure, the reaction solution is saturated with nitrous oxide (N_2O). The following reaction takes place to convert the e_{aq}^- into hydroxyl radicals:

$$e_{aq}^- + N_2O + H^+ \rightarrow N_2 + {}^{\bullet}OH$$

- If superoxide radicals are needed, then before the pulse exposure, the reaction solution is saturated with O_2 and contains sodium formate (HCOONa). The following reactions take place to give superoxide radicals:

$$e_{aq}^- + O_2 \rightarrow OO_2^{\bullet-}$$
$$H^{\bullet} + HCOO^- \rightarrow H_2 + CO_2^{\bullet-}$$
$${}^{\bullet}OH + HCOO^- \rightarrow H_2O + CO_2^{\bullet-}$$

The carbon dioxide radical is a powerful reducing agent, and it reduces oxygen to superoxide.

$$CO_2^{\bullet-} + O_2 \rightarrow O_2^{\bullet-} + CO_2$$

Since the reactions of interest are usually of second order, we need concentrations of two species — the A-derived species we mentioned earlier and the other species, say hydroxyl. As seen earlier, the A-derived species concentrations can be measured through electron paramagnetic resonance (EPR)–, fluorescence–, or absorbance–spectroscopy. The hydroxyl species concentrations are "handled" through competition methods, which use a "probe" that reacts with ${}^{\bullet}OH$. The probe is present in the same reaction mixture and hence "sees" the same concentration of ${}^{\bullet}OH$ as the species of interest, A. The product of the reaction between ${}^{\bullet}OH$ and the "probe" could be measurable, or the balance can be manipulated to yield the second-order rate constant.

For example, thiocyanate ion is used as a probe. When it reacts with $^{\bullet}OH$ radical,

$$^{\bullet}OH + SCN^- \rightarrow OH^- + SCN^{\bullet}$$
$$SCN^{\bullet} + SCN^- \rightarrow (SCN)_2^{\bullet-}$$

$(SCN)_2^{\bullet-}$ strongly absorbs at around 500 nm, so its concentration can be followed. The concentration of $^{\bullet}OH$ in the reaction mixture can be followed by stoichiometry.

A better accuracy results if the degradation of $^{\bullet}OH$ species is also considered.

2.2.5.2. *Stopped flow*

When the rates are too slow for pulse radiolysis measurement but too fast to be measured by standard biochemical methods (ms range), stopped flow is used. The reactant solutions are held in different syringes (typically two) that lead to a reactant cell made of quartz and mixed at time zero. One contains the radical, say superoxide, and the other the compound of interest. Upon mixing, the various concentrations are followed the same way as earlier through absorbance, fluorescence, or EPR spectroscopy. This has been used to study reactions of hypochlorous acid and peroxynitrite.

Additional Information

Videos:
How ultrasound works: https://www.youtube.com/watch?v=I1Bdp2tMFsY
Phacoemulsification cataract surgery: https://www.youtube.com/watch?v=MqwyoXBwFSI
Removal of kidney stones: SWL: https://www.youtube.com/watch?v=0hKRYVrlfdI
The control of blood pH: https://www.youtube.com/watch?v=aexUQAtp9WE

References

Buettner, G. R. (1993). The pecking order of free radicals and antioxidants: Lipid peroxidation, a-tocopherol and ascorbate. *Arch. Biochem. Biophys.*, 300, 525.

Halliwell, B., and Gutteridge, J. M. C. (2015) *Free Radicals in Biology and Medicine*, Ed. 5, Oxford Univ. Press, Oxford.

Loach, P. A. (1976) Oxidation-reduction potentials, absorbance bands and molar absorbance of compounds used in biochemical studies. In *Handbook of Biochemistry*, Ed. 3, (Fasman, G.D., ed.), CRC Press, Boca Raton.

Chapter 3

Measurement of Reactive Species Levels in Biological Systems

Measurement of intracellular reactive species (RS) levels/concentrations is important for better understanding the RS effects on cells. It is good to recall from the earlier chapters that the RS reaction rates are high, and thus, the actual RS concentrations change rapidly in, say, the intracellular space. Therefore, there needs to be clarity on the measured quantity because any measurement has a finite time associated with it.

Recall from Chapter 1 that we are interested in measuring the ***pseudo-steady-state (PSS) concentration*** of intracellular RS for most biological needs. However, in some chemical reaction studies, say, for determining the rate constant of a reaction, one may be interested in measuring the actual (volumetric) concentrations of RS.

Let us also be clear about the distinction between the two terms, ***concentration*** and ***specific level***. Concentration usually refers to the amount of the species in a particular volume (volumetric concentration). In other words, the species' amount (mass or moles) is normalized (mathematically, divided) with the volume of interest. The amount of the species could be normalized with respect to other measures, such as the area or mass of a containing substance/matrix. If we use such normalizations, we will explicitly mention it in this book at the point of use. Thus, unless otherwise mentioned, we will work with volumetric concentrations in this book.

The methods available for measurement typically measure the volumetric concentrations of the species. However, the experimentally obtained concentration measurement using a culture of cells could refer to the amount of species present in all the cells. The interest often is in the amount of the species present inside each cell, that is, the *specific level* of the species. Thus, the measured volumetric concentrations are normalized with cell concentrations to obtain the amount per cell or amount per mass of cells. In other words,

$$species\ concentration = \frac{species\ amount}{measurement\ volume}$$

$$cell\ concentration = \frac{cell\ mass}{measurement\ volume}$$

$$specific\ level\ of\ the\ species = \frac{species\ amount}{cell\ mass}$$

$$= \frac{(species\ amount/measurement\ volume)}{(cell\ mass/measurement\ volume)}$$

$$= \frac{species\ concentration}{cell\ concentration}$$

Some of the commonly used methods for free radical detection and quantitative measurement are as follows:

- Reaction between RS and other compound(s) to yield products that are detectable through fluorescence, absorbance, electrical measurements (through a probe), or separable by GC, HPLC, and so on, and subsequent measurement of those products.
- Electron paramagnetic resonance (EPR) or electron spin resonance (ESR) — we will use these terms interchangeably in this course. RS with unpaired electrons, that is, free radicals (FR), can be measured through EPR.

Further, let us differentiate between methods suitable for measuring intracellular RS in cultivated cells and other systems; the other systems could be in vitro or in whole bodies. For cultivated cells, fluorescence methods seem most suitable, followed by EPR, which poses difficulties in interpretation when many radicals are present together.

Next, let us see a method or two for the measurement of each RS in some detail, followed by some methods reported in the literature for their measurement. Then, we will discuss the EPR method, sometimes considered the gold standard for simple measurements, in some detail.

3.1. Hydroxyl Radicals

Hydroxyl radical measurement using a fluorescence method:

The hydroxyl radical can be determined using the nonfluorescent dyes aminophenyl fluorescein (APF) or hydroxyphenyl fluorescein (HPF).

Principle

The dyes are transformed to their fluorescent form upon reaction with hydroxyl radicals, as shown in the reaction in Figure 3.1. The species on the LHS could be either APF or HPF, depending on whether the group "A" is a (–NH) or a (–O), respectively.

The almost nonfluorescent HPF and APF are *O*-dearylated upon reaction with the hydroxyl radicals to yield strongly fluorescent fluorescein.

Calibration curve

Fenton reaction between H_2O_2 and Fe^{2+} generates stoichiometric amounts of $OH^\bullet$, which can be used to generate the standard curve, the relationship between flourescence intensities and concentration.

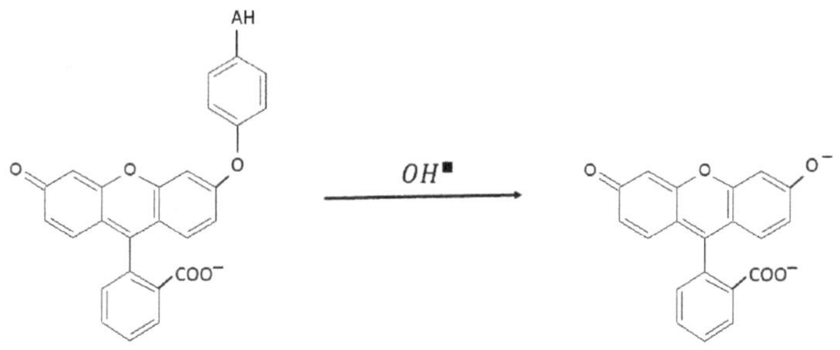

Figure 3.1. APF or HPF reaction with hydroxyl radicals

Procedure
It involves incubating the sample with the dye and measuring the resultant fluorescent intensity.

Interferences
Hypochlorite significantly interferes with APF fluorescence but not so much with HPF fluorescence (data in Setsukinai *et al.*, [2003]). We have also verified the interference extent in our laboratory. A possible strategy is to use HPF for mammalian cells because they produce hypochlorite; so APF cannot be reliably used in this case. For bacterial cells, either dye can be used. Also, peroxynitrite interferes to a certain extent.

Many methods have been reported in the literature for hydroxyl radical detection and measurement. Some of them are

– Hydroxylation of terephthalic acid
– Hydroxylation of coumarin
– Hydroxylation of phthalic hydrazide
– Hydroxylation of phenethyl polyacrylate
– Hydroxylation of antipyrine
– Hydroxylation of benzoic acid.

For details on the above methods for hydroxyl radical detection and measurement, the reader is referred to information-intensive texts such as Halliwell and Gutteridge, [2015]. Each has advantages and disadvantages, including the APF/HPF fluorescence methods.

3.2. Superoxide

Superoxide radical measurements in cell culture are often done with the fluorescent probe hydroethidine (HE), more commonly known as dihydroethidium (DHE).

Superoxide ($O_2^{\bullet-}$) reacts with the dye to form a fluorescent product 2-hydroxyethidium (2-OH-E$^+$). The formation of this product has been described as a two-stage process:

(i) The first step involves the formation of an HE radical or radical cation from a hydrogen or electron abstraction reaction

$$ HE \xrightarrow{\;O_2^{\bullet-}\;} HE^{\bullet+} \xrightarrow{\;O_2^{\bullet-}\;} 2-OH-E^+ $$

More commonly
known as DHE
(dihydroethidium)

Figure 3.2. The details of the superoxide measurement with DHE

(ii) Recombination of this intermediate radical with a second molecule of $O_2^{\bullet-}$ to form 2-OH-E$^+$. This intercalates with DNA to produce fluorescence (note: DNA is needed for the assay to work, especially while preparing the calibration curve)

Thus, two molecules of $O_2^{\bullet-}$ result in the generation of one molecule of 2-OH-E$^+$. The first stage involving the reaction between $O_2^{\bullet-}$ and DHE to the form HE radical is a relatively slow process (k = 10^3 M^{-1} s^{-1}), whereas the second step is rapid (k = 10^9 M^{-1} s^{-1}).

The calibration curve is prepared by generating different concentrations of superoxide radicals and incubating the same with the dye (final concentration of 10 μM) at 37°C for 30 minutes, along with enough DNA. The fluorescent intensity of the mixture is then measured at excitation and emission wavelengths of 405 and 570 nm, respectively.

Superoxide generation is effected via the xanthine/xanthine oxidase system. Various concentrations of xanthine are oxidized by XO to generate different concentrations of $O_2^{\bullet-}$ as per the following reaction

$$ XA + 2O_2 + H_2O \xrightarrow{\;xanthine\,oxidase\;} uric\ acid + 2O_2^{\bullet-} + 2H^+ $$

Drawbacks
The formation of the HE radical (HE$^{\bullet+}$) can be brought about by other oxidants (peroxidases and one electron oxidants) as well. However, the conversion of the radical to 2-OH-E$^+$ can be brought about only by reaction with $O_2^{\bullet-}$. Also, nonspecific products such as ethidium (E$^+$) can interfere with the fluorescence signal from the desired product, 2-OH-E$^+$.

The desired fluorescent product, 2-OH-E$^+$, has an excitation and emission maxima of 480 and 580 nm, respectively (480/580 nm). However, at 480/580 nm, the fluorescence of the nonspecific product, ethidium, significantly interferes with the measurement. To avoid the interference, an excitation/emission wavelength pair of 405/570 nm is used [Nazarewicz, 2012]. At this wavelength, the ratio of fluorescence intensities due to the desired product, 2-OH-E$^+$, and the undesired product, ethidium, is ideal. However, the absolute value of fluorescence intensity at 405/570 nm of the desired product, 2-OH-E$^+$, is considerably lower than that at 480/580 nm.

Superoxide can be measured by other methods. Some of them are

- Reduction of cytochrome C
- Reduction of nitroblue tetrazolium
- Light emission by luminol
- Light emission of luciferin
- Oxidation of adrenalin
- Oxidation of NADH + lactate dehydrogenase
- Oxygen uptake by sulfite.

For details on the above methods, the reader is referred to information-intensive texts such as Halliwell and Gutteridge, [2015].

3.3. Hydrogen Peroxide

Of the various methods available to measure H_2O_2, the fluorimetric assay using Amplex red, being extremely sensitive, is described here. H_2O_2 concentrations as low as 50 nM can be detected using this method.

Principle
Hydrogen peroxide reacts with Amplex red [N-acetyl-3,7- dihydroxyphenoxazine], a colorless and nonfluorescent derivative of resorufin at a stoichiometry of 1:1 in a reaction catalyzed by horseradish peroxidase (HRP) to generate the highly fluorescent product, resorufin, via the formation of an Amplex red radical intermediate. HRP catalyzes the decomposition of H_2O_2 to the hydroxyl radical, which is then reduced to water due to the irreversible chemical oxidation of Amplex red, thereby generating fluorescent resorufin, as shown in Figure 3.3.

The resorufin so formed is measured for its fluorescence at excitation and emission wavelengths of 563 and 587 nm, respectively.

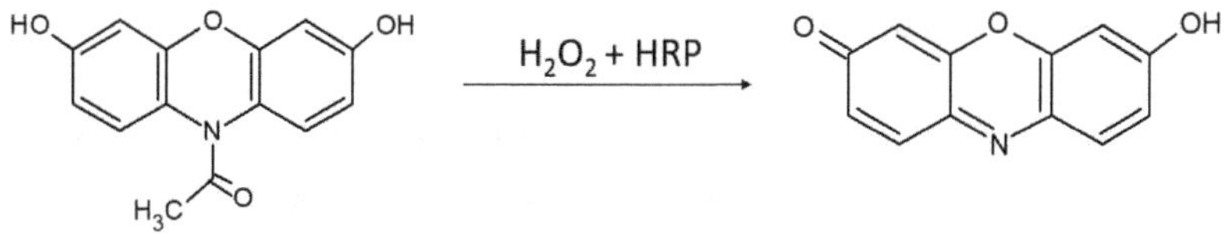

Figure 3.3.   Amplex red reaction for H_2O_2 measurement

Procedure

Different concentrations of H_2O_2 are incubated with equimolar concentrations of Amplex red in 50 mM Tris-HCl (pH 7.4) containing HRP (1 U/mL) at room temperature for 5 minutes. The fluorescence is measured under the above-mentioned excitation/emission conditions, and the calibration chart is obtained by plotting H_2O_2 concentration versus the fluorescent intensities.

Drawbacks

Photochemical oxidation of resorufin occurs in the presence of GSH, NAD(P)H, or ascorbate, thereby interfering with the assay.

Amplex red-derived radicals can rapidly react with $O_2^{\cdot-}$ and inhibit product formation.

For other methods of hydrogen peroxide measurement, the reader is referred to information-intensive texts such as Halliwell and Gutteridge, [2015].

3.4. Use of DCFHDA

At the time of this writing, a fluorescent probe for "ROS" called 2′,7′-dichlorodihydrofluorescein diacetate (DCFHDA) is highly popular. DCFHDA enters the cell and is converted to DCFH by esterases. The "ROS" convert DCFH to DCF, which is fluorescent with an optimal wavelength pair of 488/525 nm. However, it is still not completely clear which RS species DCFHDA measures. A partial list includes

$$RO_2^{\cdot}, RO^{\cdot}, NO_2^{\cdot}, RS^{\cdot}, CO_3^{\cdot-}, {}^{\cdot}OH, ONOO^{-}$$

Other complications with DCFHDA measurements are detailed in Halliwell and Gutteridge [2015]. Thus, careful consideration is needed before using DCFHDA and interpreting its fluorescence value for RS measurement.

Further, the challenges with using the fluorescence-based methods, in general, have been extensively studied [Cohn, *et al.*, 2009; Kalyanaraman *et al.*, 2012; Michalski *et al.*, 2014; Rhee *et al.*, 2010; Zielonkaa *et al.*, 2009]. A user of the techniques needs to be aware of the specific challenges.

3.5. Nitric Oxide

The chemiluminescence method is considered to be the most sensitive method. The method is sensitive up to 10^{-13} M of NO.

Principle
The method is based on the measurement of luminescence produced due to a chemical reaction [Archer, 1993].

Ozone interacts with NO to generate light, as shown below

$$NO^{\bullet} + O_3 \rightarrow NO_2^* + O_2$$
$$NO_2^* \rightarrow NO_2 + light$$

The luminescence thus produced is proportional to the concentration of NO and is measured with a sensitive photomultiplier tube (PMT).

Calibration curve
The PMT is calibrated using gaseous NO or aqueous solutions containing various amounts of NO (20–1000 pmol). The signal from the PMT in mV is plotted against the concerned concentration of NO to obtain the calibration curve. The calibration is carried out with gaseous NO or NO in solution based on the state of the sample of interest.

Gaseous NO is commercially available. Different concentrations of the same may be obtained by serial dilution. In the case of a liquid sample, dilutions of a saturated NO solution (3 mM NO) using deoxygenated water are prepared.

Preparation of saturated NO solution
Double-distilled cold water is bubbled with helium for 30 min to remove O_2. This water is then bubbled with pure NO (>99.0% NO) for 30 min in

a glass sampling bulb. Standards are made fresh daily and kept in glass bulbs that can be aspirated through a rubber septum with a gastight syringe.

Procedure

It involves mixing ozone (generated internally by electrical discharge) with gaseous NO in a reaction chamber in front of a cooled, red-sensitive PMT. In the case of aqueous specimens, NO is stripped from the same by bubbling the sample with inert gas under vacuum. This is possible due to the low solubility of NO in aqueous solutions. The signal from the PMT is then recorded and compared with the calibration curve.

Drawbacks

NO is rapidly destroyed by oxygen.

Precaution must be exercised while dealing with biological samples as these tend to be proteinaceous and thus foam during stripping and aspiration into the reaction chamber, thereby coating the PMT and reducing its sensitivity.

3.6. Electron Paramagnetic/Spin Resonance (EPR or ESR) Spectroscopy

EPR spectroscopy was discovered in 1945 by Zavoisky, and within 10 years, biological samples were studied using that technique. It measures the paramagnetism due to the unpaired electrons. The unpaired electron has a spin, which has an associated magnetic moment. When an external magnetic field of field strength, H, is applied, the unpaired electron can align itself such that its spin magnetic dipole moment is aligned with or against the applied external field. The energy states associated with these two types of alignments (with or against) are different. The alignment "with" the applied field, that is, parallel orientation, has lower energy than the alignment "against" the applied field, that is, antiparallel orientation. The difference in energy levels between these two energy states is

$$\Delta E = g\,\beta H$$

where

g = spectroscopic splitting constant (=2.002319278 for an electron; a universal physical constant)

H = applied magnetic field strength

β = Bohr magneton (a constant) = 9.274×10^{-24} JT^{-1} $particle^{-1}$ *here,* $T = Tesla$

For most lab measurements, $\Delta E \approx 10^{-3}$ $kcal\ mol^{-1}$, which is insignificant compared to the energies associated with chemical reactions.

When an energy equal to the difference between the energies of the two orientations, that is, ΔE, is provided by any means, resonance sets in, and transition from the lower to the higher energy states occurs with net energy absorption. This energy can be measured, and this measurement provides the basis for the EPR/ESR spectroscopy. If the subscript, r, is used to represent resonance conditions, then,

$$\Delta E = h v_r = g\,\beta H_r$$

$$\Rightarrow \frac{v_r}{H_r} = \frac{g\beta}{h} = constant$$

Thus, the ratio, $\frac{v_r}{H_r}$ is relevant and not the actual value of v_r or H_r. Therefore, in principle, either v_r or H_r can be varied to provide the resonance conditions. All EPR spectrometers operate at fixed v_r and vary H_r.

The energy absorption spectrum and the more useful first-derivative spectrum are illustrated in Figure 3.4.

Magnetic fields of up to 15,000 G are usually obtainable under laboratory conditions. Thus, the corresponding frequencies are <42 GHz, corresponding to the microwave region (Zavoisky used Klystron tubes). Typically, frequencies in the 9–10 GHz range are used. Based on the frequency range, there are different band spectrometers available: S-band (3 GHz), X-band (9.5 GHz), K-band (23 GHz), Q-band (35 GHz) and W-band (95 GHz).

Note that the resonance condition described above took into account only the spin magnetic moment, which arises due to the spinning of an electron. Other contributions could be significant:

(1) An orbital angular momentum due to the motion of the electron in a "circular path."

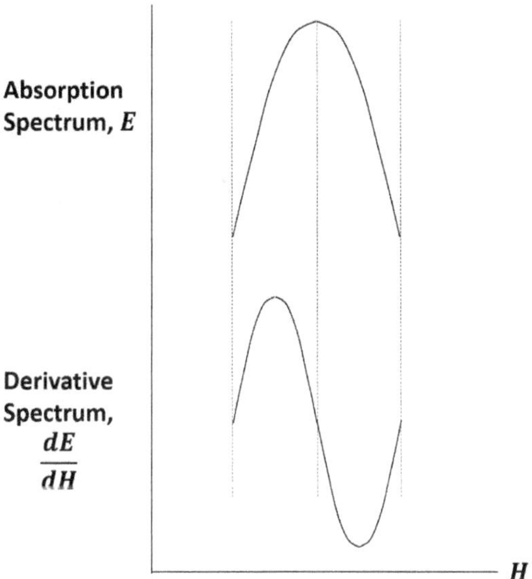

Figure 3.4. Typical energy absorption spectrum and its derivative spectrum with respect to the applied magnetic field strength

(2) Interactions with nuclear magnetic moments, if present—the nuclear hyperfine interactions.

The first aspect is handled using an "effective g value" or "g factor," which characterizes the molecule in which the electrons are located. Usually, no subscript is used to denote this "g factor."

$$g_{eff} = g\ factor,\ or\ g = \frac{h\nu_r}{\beta H_r}$$

The typical g factor varies from 2.0024 to 2.06 depending on the class of molecules. A table with the details of the various typical g-values in each class of molecules is available in Borg [1983].

The second aspect of interactions with nuclear magnetic moments is more involved. The energy levels (E) of the parallel ($m_s = -1/2$) and anti-parallel ($m_s = +1/2$) orientations/spins of electrons as a function of the externally applied magnetic field strength (H) can be represented as given in Figure 3.5. At the resonance value, the energy of the applied field

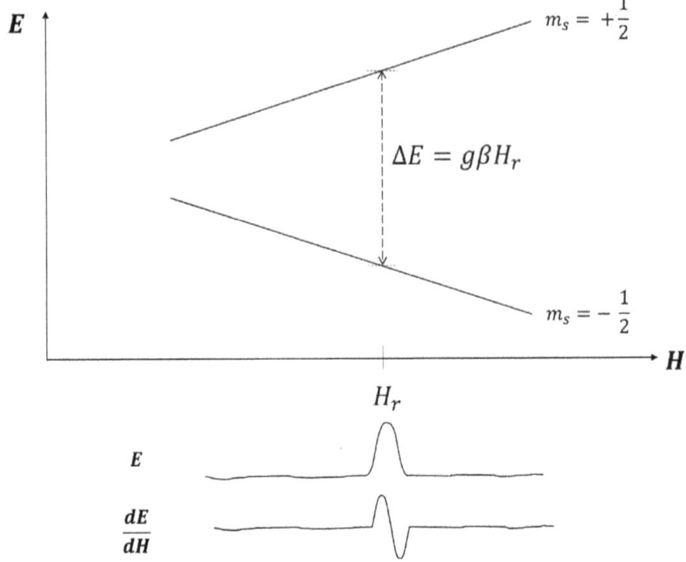

Figure 3.5.   The energy levels of the parallel and antiparallel orientations/spins of electrons in an applied magnetic field

matches the energy level difference between the two spin orientations, and therefore, the relevant species containing those electrons absorbs energy.

Now, let us introduce a nucleus (a proton) into the picture to understand the first-level interactions. Each nucleus of spin I causes the following split in the energy levels, that is, it brings into existence more energy levels.

The nuclear spin interactions cause more electron energy levels of the same energy difference to become available for transition. In other words, they "split" the electron energy levels. The energy level splits, usually followed through the derivative spectra, are referred to as hyperfine splits. Each nucleus of spin I splits the electron energy levels into (2I + 1) sublevels. However, note that ^{12}C and ^{16}O do not possess a net nuclear magnetic moment. Therefore, the nuclei of ^{12}C and ^{16}O will not split the electron energy levels. On the other hand, ^{1}H, ^{2}D, ^{14}N, ^{13}C, and other nuclei with a net nuclear magnetic moment will split the electron energy levels.

The split in the energy levels with one nucleus (Figure 3.6) can also be represented as a stick diagram. Stick diagrams are simpler and allow

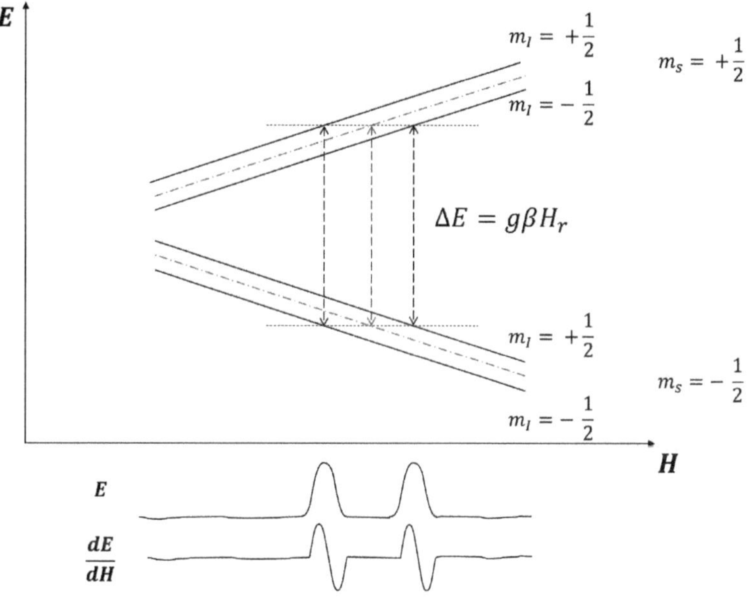

Figure 3.6.   The split in energy levels with one nucleus

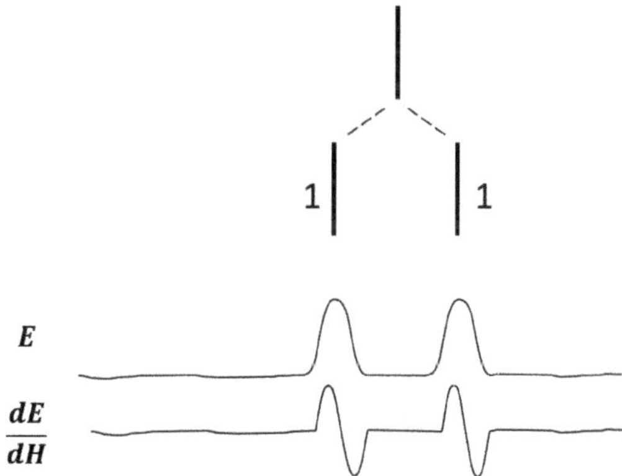

Figure 3.7.   Stick diagram of the split in electron energy levels with one nucleus

easier scale-up to represent the splits when multiple nuclei are considered. The stick diagram for the split in the energy levels with one nucleus is presented in Figure 3.7.

Figure 3.7. represents the spectra that hydroxyl radicals ($OH^{\bullet}$) provide when present alone. Now, if the interaction of two equivalent protons

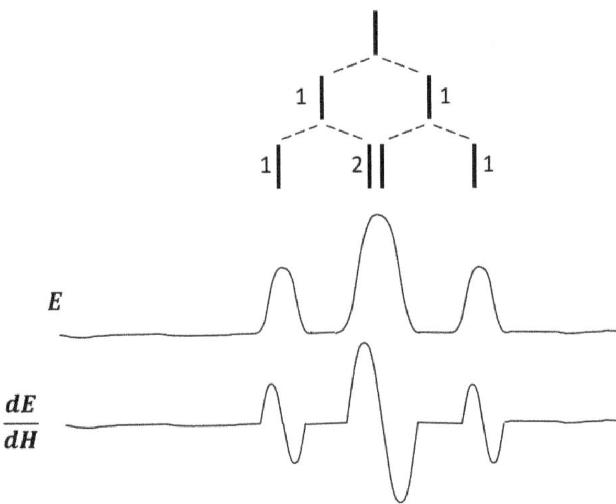

Figure 3.8.　Interaction of two equivalent protons

is considered, the energy level split is illustrated as a stick diagram in Figure 3.8.

The stick can be easily extended to represent the splits when more than two equivalent protons are in the picture. Figure 3.9 represents two extensions.

In general, n equivalent protons give a spectrum of $(n + 1)$ hyperfine splits. Their intensities are given by the binomial expression ($I(M)$ represents the energy absorbed by the Mth line)

$$I(M) = \frac{n!}{\left(\dfrac{n}{2} + M\right)!\left(\dfrac{n}{2} - M\right)!}$$

$$M = -\frac{n}{2}, -\frac{n}{2} + 1, \ldots, \frac{n}{2} - 1, \frac{n}{2}$$

Useful characteristics of the EPR spectrum:

－ The value of the applied magnetic field, H, at which resonance occurs, which is determined by the g factor

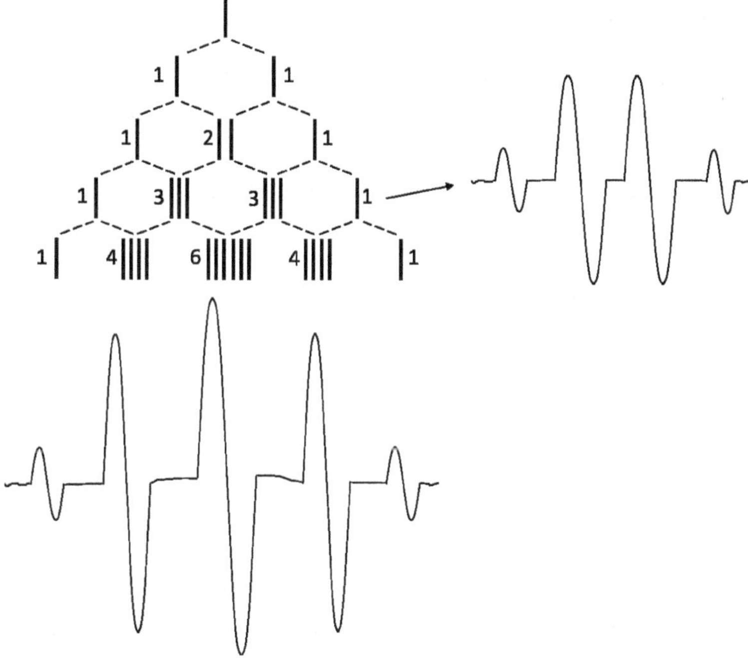

Figure 3.9. Stick diagram with multiple equivalent protons

– Area under the absorption curve that indicates the concentration of the RS

– The number of spectral lines, separation of lines, width and shape of individual lines (we have not covered these aspects)

Spin traps are often used to better capture the radical through EPR. The spin traps interact with the radicals and form reasonably stable FR. The stability can last for a few hours, and thus, measurement becomes easier with their use. For example, 5,5-di-methylpyrolline-N-oxide (DMPO) is a spin trap for hydroxyl radicals. The reader is referred to information-intensive texts (e.g., Halliwell and Gutteridge, [2015]) for the details on spin traps, their spectra, and their use. Note that the spectrum of spin-trapped hydroxyl radical will not be the same as the native hydroxyl radical, but it will be distinct if hydroxyl alone is present to enable its easy identification.

3.7. Biomarkers

Some investigators, especially in medicine, are interested in measuring the outcomes of the interactions of RS with cells or the various biomolecules.

The products of the reactions between RS and biomolecules can be considered as fingerprints of the damage if

- They are unique
- Not generated by normal biological processes
- Do not originate from diet.

Damage to DNA, protein, or lipid is estimated in various ways in samples drawn from the subjects. Lipid peroxidation by the popular ThioBarbituric Acid Reaction Species (TBARS) assay has been used in many fields, starting with food preservation, as a marker of lipid damage. However, the TBARS assay, which is simple to perform, has significant limitations in reflecting the realities of lipid peroxidation. The reader is referred to information-intensive texts such as Halliwell and Gutteridge [2015] for the complications involved in interpreting TBARS assay.

One of the desired features of a biomarker is its availability in easily accessible sources such as urine and other bodily fluids. Researchers have used urate oxidation products such as allantoin as a damage marker. The more the damage, the more the allantoin levels in urine, wound fluid, microdialysis fluid, and so on.

However, as discussed in an earlier chapter, biomarkers provide information about the damage that has already occurred in a biological system. Thus, they have a diagnostic value and highly limited, if any, therapeutic value.

References

Archer, S. (1993). Measurement of nitric oxide in biological models. *FASEB J.*, 7, 349.

Borg, D. C, (1983) Applications of ESR in Biology, *Free Radicals in Biology*, Ed. Pryor W. A., Academic Press, New York.

Cohn, C. A., Pedigo, C. E., Hylton, S. N., Simon, S. R., and Schoonen, M. A. (2009). Evaluating the use of 3'-(p-Aminophenyl) fluorescein for determining the formation of highly reactive oxygen species in particle suspensions. *Geochem. Trans.*, 10, 8.

Halliwell, B., and Gutteridge, J. M. C. (2015) *Free Radicals in Biology and Medicine*, Ed. 5, Oxford Univ. Press, Oxford.

Kalyanaraman, B., Usmar, V., Davies, K. J. A., Dennery, P. A., Forman, H. J,. Grisham, M. B., Mann, G. E., Moore, K., Roberts, L. J., and Ischiropoulos, H. (2012). Measuring reactive oxygen and nitrogen species with fluorescent probes: challenges and limitations. *Free Radic. Biol. Med.*, 52, 1.

Michalski, R., Michalowski, B., Sikora, A., Zielonka, J., and Kalyanaraman, B. (2014). On the use of fluorescence lifetime imaging and dihydroethidium to detect superoxide in intact animals and exvivo tissues: a reassessment. *Free Radic. Biol. Med.*, 67, 278.

Nazarewicz, R. R., Bikineyeva, A., and Dikalov, S. (2012). Rapid and specific measurements of superoxide using fluorescence spectroscopy. *J Biomol. Screen.*, 18, 498.

Rhee, S. G., Chang, T., Jeong, W., and Kang, D. (2010). Methods for detection and measurement of Hydrogen Peroxide inside and outside of cells. *Mol. Cells*, 29, 539.

Setsukinai, K., Urano, Y., Kakinuma, K., Majima, H. J., and Nagano, T. (2003). Development of novel fluorescence probes that can reliably detect reactive oxygen species and distinguish specific species. *J. Biol. Chem.*, 278, 3170.

Zielonkaa, J., Hardy, M., and Kalyanaraman, B. (2009). HPLC study of oxidation products of hydroethidine in chemical and biological systems: ramifications in superoxide measurements, *Free Radic. Biol. Med.*, 46, 329.

Chapter 4

Reactive Species in Cell Signaling

Thus far, we have seen the "reactive" and, hence, the potentially deleterious effects of reactive species (RS). RS serve many physiological roles in the cell. Let us look at the signaling role played by RS in this chapter. Many applications of RS in manipulating biological systems toward desired ends, given in the later chapters, are based on the signaling aspects and sublethal effects of RS.

4.1. Fundamental Mechanisms

The usual means by which RS effect cell signaling processes is through chemical modifications of relevant proteins. The common modifications occur on proteins [D'Autreaux and Toledano, 2007; Halliwell and Gutteridge, 2015] at

1. Fe–S clusters (centers): Oxidation of Fe^{2+} followed by reduction to cycle back to the original form.
2. –SH group: Oxidation/reduction of either a single –SH group or a pair of –SH groups to form disulfide, followed by reduction to cycle back to the original form.
3. Histidine groups: Oxidation to form oxo-histidine followed by reduction to cycle back to the original form.

The above oxidation/reduction processes could influence cell signaling through two means:

1. By directly influencing the relevant protein function (e.g., p53 in mammalian cells)
2. By influencing the transcription factor function, and thus the production (at the transcription level) of various relevant proteins (e.g., SoxR, OxyR in bacteria)

These actions at the molecular level translate to action at the metabolic pathway level or genetic pathway levels and ultimately to action at the cell level. Let us look at some examples, first in bacteria and then in mammalian cells.

4.2. In Bacteria

In bacteria, the **SoxR** protein performs its signaling action through a redox mechanism. The SoxR protein functions as a transcription factor. It has an iron–sulfur center in the $(2Fe - 2S)^+$ state when inactive (Figure 4.1). When the $(2Fe - 2S)^+$ centers are oxidized by, say, $O_2^{\bullet-}$, to $(2Fe - 2S)^{2+}$, the activated SoxR acts as the transcriptional activator for the *soxS* gene to result in the production of SoxS protein. The SoxS protein activates the production of about a hundred other proteins that respond to increased RS levels, for example, some SODs, a DNA repair enzyme (endonuclease IV), and others.

To get back the inactive form of SoxR, NADPH (Figure 4.1) reduces the activated SoxR through a series of electron transfers. The redox cycle continues.

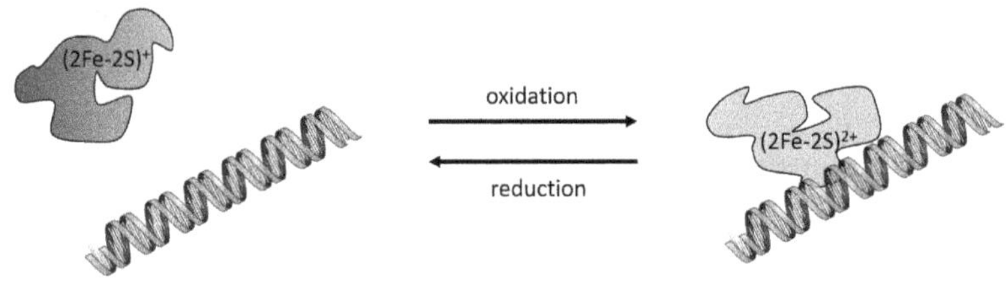

Figure 4.1.   Activation of SoxR through redox action

An example of the –SH group modification by redox action in **OxyR** to effect cell signaling is shown in Figure 4.2. Let us recall that cysteine (cys) has a –SH group. The cys residue in position 199 of the transcription factor, OxyR, can be oxidized by, say, H_2O_2. The oxidation leads to an intramolecular di-sulfide bond between cys 199 and cys 208. The di-sulfide bond formation leads to a conformation change in OxyR, which enables its binding to the DNA followed by transcription of the relevant genes. The cycle is completed by the reduction of the activated OxyR by glutaredoxin-1, and the redox cycle continues.

The third example of redox signaling in bacteria after SoxR and OxyR that we will consider is the oxidation of the cys–zinc redox center of the **Hsp33** chaperone protein (Figure 4.3). Oxidation of cys–zinc redox center by, say, (heat + H_2O_2), or HOCl causes the formation of two disulfide bonds with zinc release. The resulting conformational change activates Hsp33.

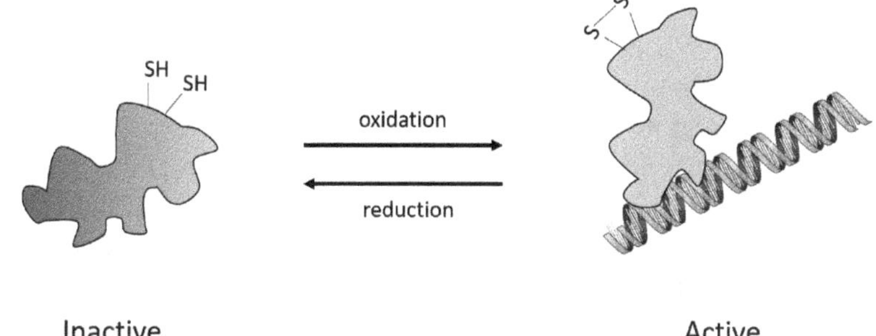

Figure 4.2.   Activation of OxyR through redox action

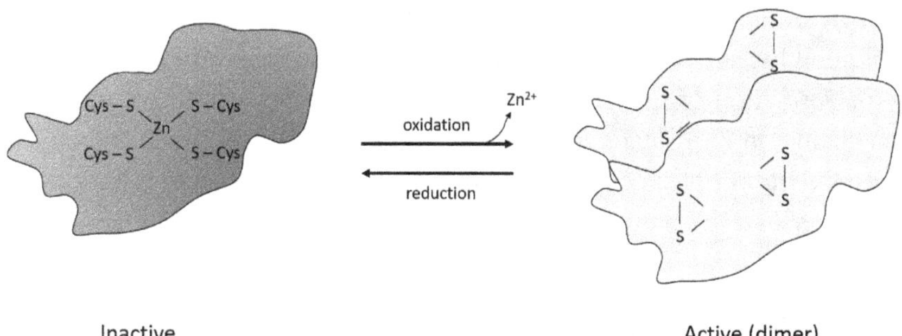

Figure 4.3.   Activation of Hsp33 through redox action

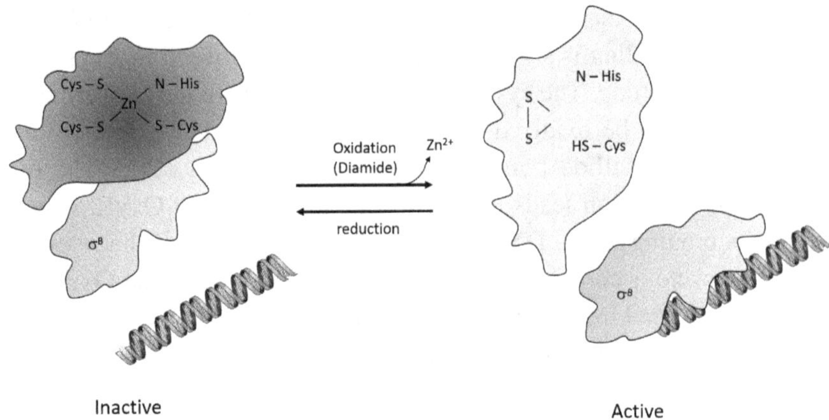

Figure 4.4.   Activation of transcription through σ^B-DNA binding through redox action on RsrA

A similar activation mechanism is seen in the context of the action of the anti-sigma factor, RsrA (Figure 4.4). The σ^B is a transcription factor that binds to DNA and promotes the specific binding of RNA polymerase to the promoter region of a gene. Typically, the binding between σ^B and RsrA prevents σ^B from binding to the DNA. The oxidation of the cys–zinc redox center in RsrA by, say, diamide leads to the formation of one di-sulfide bond with Zn release. This oxidation results in the loss of binding between σ^B and RsrA, and thus σ^B is free to bind to DNA to act as a transcription factor.

The **PerR** protein functions as a transcription factor — it acts as a functional orthologue of OxyR. PerR is also involved in redox signaling through a mechanism that is a variant of the ones described above. Oxidation by, say, the hydroxyl radical from H_2O_2 releases Fe^{2+} and leads to the formation of oxo-histidines (Figure 4.5). The resulting conformation change in PerR leads to the loss of bonding between PerR and DNA. Thus, the transcription of the relevant protein, which was repressed earlier by the PerR–DNA binding, gets derepressed.

4.3. In Mammalian Cells

Now, let us turn our attention to mammalian systems. Mitogen-activated protein kinase (MAPK) kinase cascades play central roles in responses to

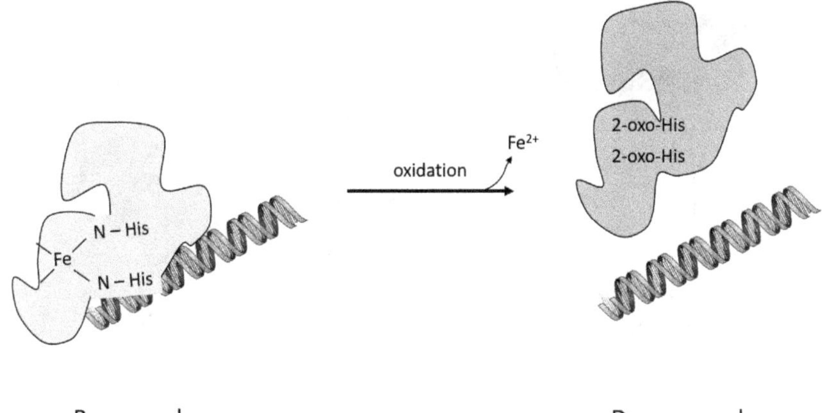

Figure 4.5.   Oxidation of PerR by hydroxyl radicals from H_2O_2

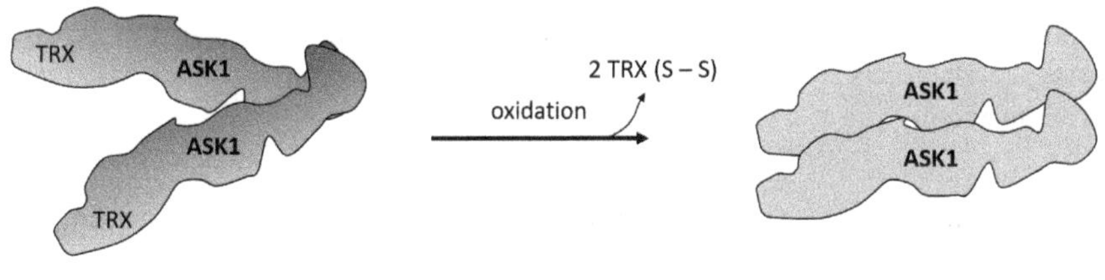

Figure 4.6.   Oxidation by ROS directly activates ASK1

stresses, growth factors, hormones, cytokines, and so on, in mammalian cells. These important cascades consist of three types of kinases, namely:

1.   Extracellular-signal Related Kinases (ERK)
2.   c-Jun N-terminal Kinases (JNK), and
3.   p38 Kinases (p38).

The principle is that phosphorylation (addition of a phosphate group) activates the relevant proteins, followed by dephosphorylation to complete the cycle. The activation happens as follows: MAPKKK-type proteins activate MAPKK-type proteins; MAPKK activates MAPK. For example (Figure 4.6), Apoptosis Signal-regulated Kinase 1 (ASK1) is a MAPKKK

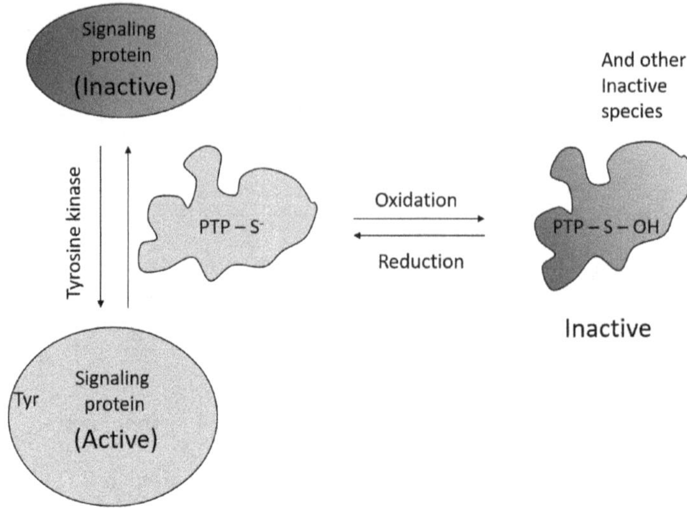

Figure 4.7.   Oxidation of PTP by RS to its sulfenic acid, sulfonic acid, and other inactive forms jams the reverse reaction for the active form of the signaling protein to convert into its inactive form

that regulates JNK and p38. ASK1 is ***directly activated*** by ROS induced by stress.

The ROS action could be indirect, too. In the context of the action of growth factors, cytokines and hormones, some signaling proteins are activated through phosphorylation by tyrosine kinases. Those signaling proteins are dephosphorylated by protein tyrosine phosphatase (PTP) to complete the loop (Figure 4.7). ROS inactivates dephosphorylation by oxidizing the –SH group of PTP to sulfenic acid and other species. This makes the cycle "stuck" at the active position.

There are many examples of the "indirect" role of ROS in mammalian cell signaling. Let us look at just two of them. The Phosphoinositide 3 Kinase (PI3K) pathway plays an important role in mediating the effects of growth factors and insulin. Phosphatase and tensin homology phosphatase (PTEN) is an important *negative* regulator of the PI3K pathway. If PTEN is inactive, PI3K is active. ROS (H_2O_2) oxidization inactivates PTEN, and thus it activates the PI3K pathway. The reduction of PTEN inactivates the PI3K pathway. The second example is that of the Nuclear erythroid-2 related factor 2 (Nrf2)–Antioxidant Response Element (ARE) pathway, which is an important pathway in the defense mechanism against oxidative stress. Nrf2–ARE controls the expression of genes that produce proteins, which act against oxidative toxic substances produced by ROS interactions and effect detoxification. ROS oxidation of a complex (KEAP–Cul3–Nrf2) releases Nrf2 from it. The complex is initially in the cytoplasm. Nrf2 gets into the nucleus, binds with ARE in the nucleus, which gets further activated by other oxidized proteins. ARE is a transcriptional activator for many genes with antioxidant functions. Huang *et al.* [2023] recently showed that the RS induced in the cellular environment by direct current can be coupled to the KEAP–Nrf2–ARE system to activate appropriate genes in implanted engineered human cells, which are encapsulated in a mouse, to release insulin on a need basis.

4.4. Effect on the Whole System

The above-mentioned RS effect at a molecular level on cells are translated to many effects at the systemic level, which affect the entire organism/human being. Interestingly, a Nobel Prize-winning discovery in the early part of the 20th century later turned out to be RS-based. Otto Warburg discovered in 1908 that sea urchin eggs increased their oxygen consumption by sixfold when fertilized. When this observation and related aspects were viewed in the light of cancer, it led to the Warburg effect, which won the Nobel Prize about two decades after the original discovery in 1908 [Finkel, 2011]. The Nobel prize was given to Warburg for elucidating "the nature and mode of action of the respiratory enzyme." However, almost a hundred years after the original observation, it became clear that the significant increase in oxygen consumption by sea urchin eggs after fertilization is used to produce H_2O_2 for the development of the protective shell around the fertilized egg. Thus, RS regulate essential and crucial processes in the whole organism.

RS are known to be key regulators of vascular health and diseases [Chen *et al.*, 2018]. It has been shown that RS play a critical role in hypertension, especially its chronic pathogenesis. It has also been shown that a reduction in RS reduces blood pressure. Also, there exists an "oxidative stress theory" for atherosclerosis, which says that RS oxidation of low-density lipoproteins by vascular cells leads to inflammatory responses and foam cell formation in plaques. Similarly, RS's role in restenosis and abdominal aortic aneurysm has been established. Further, the RS's roles with both positive and negative consequences in many human diseases, such as cancer, inflammatory diseases, neurological diseases, diabetes, infertility, and so on, have been established [Yang and Lian, 2020]. Thus, RS signaling plays a vital role in many physiological and pathological conditions in the human body.

References

Chen, Q., Wang, Q., Zhu, J., Xiao, Q., and Zhang, L. (2018). Reactive oxygen species: key regulators in vascular health and diseases. *British J. Pharm.*, 175, 1279.

D'Autreaux, B., and Toledano, M. B. (2007). ROS as signalling molecules: mechanisms that generate specificity in ROS homeostasis. *Nat. Rev.: Mol. Cell Biol.*, 8, 813.

Finkel, T. (2011). Signal transduction by reactive oxygen species. *J. Cell Biol.*, 194, 1.

Halliwell, B., and Gutteridge, J. M. C. (2015). *Free Radicals in Biology and Medicine*. Ed. 5, Oxford Univ. Press, Oxford.

Huang, J., Xue, S., Buchmann, P., Teixeira, A. P., and Fussengger, M. (2023). An electrogenetic interface to program mammalian gene expression by direct current. *Nat. Metab.*, 5, 1395.

Yang, S., and Lian, G. (2020). ROS and diseases: role in metabolism and energy supply. *Mol. Cell. Biochem.*, 467, 1.

Chapter 5

Reactive Species in Oxygen Supply

In the earlier chapters, we discussed the relevance of reactive species (RS), the analysis of RS reactions, the measurement of RS concentrations, all in a biological context, and the role of RS in cell signaling. In this chapter, we will begin discussing some manipulations of RS toward desired objectives from biological systems, that is, from a biological engineering perspective, and continue that discussion almost until the end of the book. To ease understanding and appreciation, let us also discuss the manipulations as stories, somewhat similar to how they unfolded in our laboratory and mathematical representation research over the past few decades.

As may be known, bioreactors are instrumented and controlled vessels that are used to produce many valuable substances of biological origin (i.e., using cells, enzyme reactions, or relevant other cell components). Bioreactors are extensively used in the bio-industry — it is the heart of any bioprocess. Most industrial bioprocesses employ aerobic organisms. Thus, oxygen is a critical nutrient that needs to be supplied to bioreactors.

Air is the most common source of oxygen to bioreactors. Air is a mixture of predominantly oxygen and nitrogen. Nitrogen in the air is usually an inert substance in the context of bioreactors. Hence, whole air is used to supply oxygen to the bioreactor after suitable filtration to remove the microorganisms present in it. In addition, enough moisture is added to the air to prevent it from stripping water from the bioreactor contents as it passes through the bioreactor.

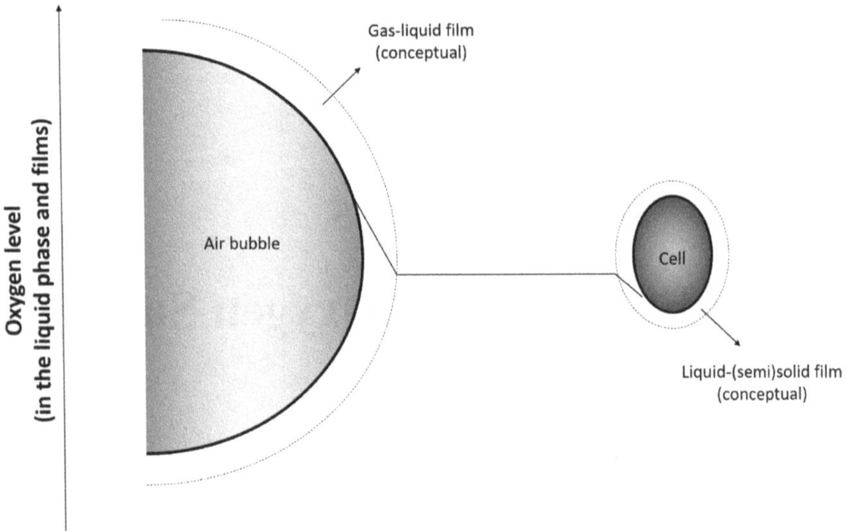

Figure 5.1.   The oxygen levels in the liquid and the conceptual films when aeration is used to supply oxygen to a suspension of single cells in a bioreactor

When air is the source of oxygen, the transport of oxygen from the air bubble to the cell, both present in the bioreactor broth, is a classic case of gas–liquid mass transport. The species oxygen moves from, say, the center of the air bubble to its surface, crosses the conceptual, gas–liquid film, moves through the liquid, crosses the conceptual, liquid–(semi-)solid film next to the cell to enter the cell (Figure 5.1). The conceptual films offer resistance to the transport of oxygen, and the gas–liquid film provides the highest resistance in systems with single, un-clumped cells. The resistances significantly decrease the rate at which oxygen can be supplied to the cells. Thus, fundamentally, the kinetics of oxygen supply is a significant challenge for effective aerobic bioreactor operation.

The supply of oxygen is also somewhat of a challenge from the other fundamental perspective — thermodynamics. The oxygen solubility in water is low at about 8 ppm under operating conditions. Thus, one cannot store oxygen in the liquid for it to be more readily available to the cells. Several approaches have been reported in the literature to increase the oxygen-holding capacity of the bioreactor broth, including oxygen vectors. However, the success of those methods has been limited. In addition, we know from the earlier chapters that storing oxygen in a cellular environment may not be advisable due to its toxicity.

5.1. The Idea

We focused on the other fundamental aspect, namely kinetics. Our thinking [Sriram *et al.*, 1998] was as follows: if the gas–liquid film resistance is the most significant resistance that impedes the oxygen transport in this system, can we somehow eliminate it? Thus, we would be fundamentally overcoming the most significant resistance to oxygen supply and can comprehensively solve the oxygen supply problem. As long as there are two phases, gas and liquid, there will be a gas–liquid film. Therefore, to eliminate the film, we need to eliminate one of the phases. The cells are in the liquid, and thus, we cannot get rid of it. Therefore, we can eliminate the gas phase. If so, we need to generate the oxygen molecule in the liquid phase itself. In other words, we need a liquid phase reaction that generates oxygen.

When we searched for a suitable reaction, we were attracted to the following one:

$$2H_2O_2 \xrightarrow{\ catalase\ } 2H_2O + O_2$$

It was attractive because it is a physiological reaction — it occurs in the normal cell to neutralize the H_2O_2 formed in an aerobic cell. All aerobes have the enzyme, catalase, to make this reaction possible. Further, it is a clean, green reaction, which produces only oxygen and water. Thus, it seemed a good candidate reaction for our purpose.

5.2. Feasibility Check

We first checked the feasibility of providing oxygen through the above liquid phase reaction. A pulse of H_2O_2 to result in its concentration of 200 μM in the bioreactor was added to 0.78 g L^{-1} of *Xanthomonas campestris* present in it (medium with cells). Upon the pulse addition, say, at time zero, the dissolved oxygen (DO) steadily increased with time (Figure 5.2).

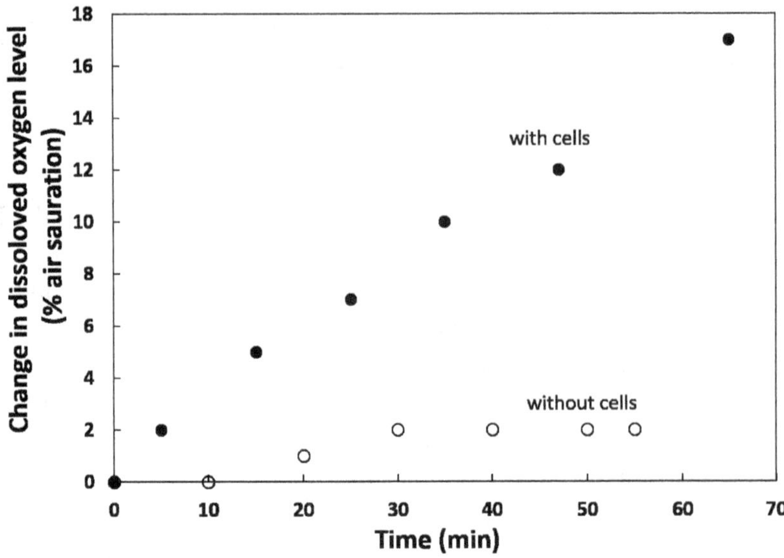

Figure 5.2.   The change in medium DO levels, with and without cells, when a pulse of 200 μM H_2O_2 was added at time zero

The increase in DO was about 18% in 60 min after the pulse addition. When the same H_2O_2 pulse was added to the medium in which *X. campestris* cells were previously grown but were removed before the addition (medium without cells), there was no significant increase in DO. This showed that it is possible to increase the DO in the medium when cells were present through H_2O_2 pulse addition. Thus, the liquid-phase oxygen-supply strategy (LPOS) is possible.

5.3. Cultivation with the LPOS

After the feasibility of providing oxygen to *X. campestris* culture with LPOS was established, the same organism was cultivated in a bioreactor with manual H_2O_2 pulse additions to supply oxygen. The DO increased immediately after a pulse of H_2O_2 and decreased after a maximum value was reached. When the DO level was close to zero, another pulse was manually added. The next pulse addition led to an increase in DO, and the cycle followed.

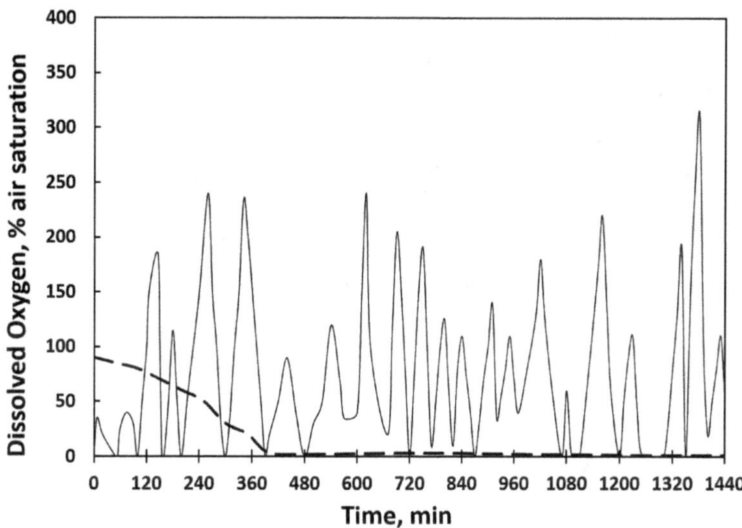

Figure 5.3. Dissolved oxygen profiles from a conventional cultivation with aeration (dashed line) and an LPOS-based cultivation (solid line) of *Xanthomonas campestris*

The DO profiles with LPOS and conventional aeration are shown in Figure 5.3. Please note that the Y axis gives percentage air saturation. Thus, complete saturation with air that contains 21% oxygen corresponds to a DO value of 100%. The theoretical maximum for the DO value is about 476% air saturation, corresponding to the equilibrium value with 100% oxygen in the gas phase. In Figure 5.3, data for the first 24 h are presented — please see Sriram *et al.* [1998] for more complete data. The DO went to 0 in about 6.5 h, and remained there until 48 h. The zero value of DO indicates that the oxygen demand rate was higher than the oxygen supply rate. In contrast, LPOS was able to maintain DO much above the target set-point of 50% throughout the 48 h period. Of course, the variation in DO due to pulse addition was a part of the strategy.

The maximum specific growth rate of the LPOS cultivation, 0.098 h^{-1}, was comparable to that obtained with the conventional cultivation, 0.11 h^{-1}. Interestingly, the maximum xanthan gum concentration with LPOS was 8.3 g L^{-1}, whereas it was only 4.91 g L^{-1} in the conventional cultivation. In other words, the volumetric production of xanthan gum with LPOS was 69% higher than with conventional cultivation.

More interestingly, the maximum cell concentration obtained in the stationary phase with LPOS was much lower than in the conventional cultivation. Thus, with LPOS, each cell was producing much more xanthan gum. On the other hand, the lower maximum cell concentration indicates the possible toxicity of H_2O_2 when used over many hours. Our desire to look closer at mitigating this toxicity motivated our entry into the world of RS and its manipulations in biological systems. It became an area for a professional lifelong engagement for me. The remaining chapters of this book address our various contributions in this field.

5.4. Mathematical Representations of Oxygen Availability

We developed a mathematical model that helped us visualize what to expect before we did the experiments. Let me present some aspects of the model here and show the results as sketches. The figures that give the actual numbers are available in Sriram *et al.* [1998].

We wrote a mass balance on the species, oxygen, over the bioreactor broth (system; strictly speaking, the system here was bioreactor broth without the air bubbles). Recall that the fundamental mass balance equation is

$$r_i - r_o + r_g - r_c = \frac{d(m)}{dt} \qquad \text{Eq. 5.4 — 1}$$

In Eq. 5.1, the "r" represents the rate, and the subscripts represent i — input, o — output, g — generation, and c — consumption, all relevant to oxygen. The RHS represents the accumulation rate of the mass of the oxygen in the defined system. The oxygen **input** that crosses the system boundary is through gas–liquid mass transport from the air bubbles to the broth (without the air bubbles). Thus

$$r_i = k_L a \left(C_{O_2}^* - C_{O_2} \right) \qquad \text{Eq. 5.4 — 2}$$

where $k_L a$ is the volumetric oxygen transfer coefficient, C_{O_2} is the oxygen concentration in the broth (dissolved oxygen), and the star superscript on it indicates saturation conditions.

The $k_L a$ was taken to vary as [Hocker *et al.*, 1981]

$$k_L a = 0.105 \frac{\dot{V}}{V} \left(\frac{P_g}{\dot{V} \rho_L \left(\frac{g\eta}{\rho_L} \right)^{0.67}} \right)^{0.6} \left(\frac{\eta}{\rho_L D_L} \right)^{-0.3} \qquad \text{Eq. 5.4 — 3}$$

where V is the broth volume, $\dot{V}$ is the volumetric gas flow rate into the bioreactor, ρ_g is the power input into the broth with dispersed gas, ρ_L is the density, and D_L is the diffusivity. The broth viscosity, η, in the above equation was found through an expression given by Herbst *et al.* [1981]

$$\eta = \eta_s \left(1 + \eta_i \, P \, exp(k_\eta \eta_i P) \right) \qquad \text{Eq. 5.4 — 4}$$

The product concentration, P, in the above equation was found through a logistic equation

$$P = P_0 + 0.869 \left(e^{\mu_p t} - 1 \right) \qquad \text{Eq. 5.4 — 5}$$

The oxygen **output** that crosses the system boundary, if present, will be through nucleation and the formation of gaseous oxygen in the system under supersaturation conditions. Such wasteful oxygen evolution is avoided by design; hence, the oxygen output rate was set to zero.

The oxygen **generation** in the system is through the decomposition of H_2O_2. Thus

$$r_g = -k_0 C_{H_2O_2} C_{catalase} \qquad \text{Eq. 5.4 — 6}$$

The basis for the rate expression above and other details are available in the paper [Sriram *et al.*, 1998].

The oxygen **consumption** is through the growth and maintenance aspects of the organisms in culture. Based on an earlier study [Pinches and Pallent, 1986], it was taken as

$$r_c = cr_x + eX \qquad \text{Eq. 5.4 — 7}$$

where r_x is the culture growth rate, X is the cell concentration, and c and e are constants. The growth rate in the logarithmic phase was taken as

$$X = X_0 e^{\mu_{max} t}$$

Eq. 5.4 — 8

where X_0 is the initial cell concentration, μ_{max} is the maximum (logarithmic phase) specific growth rate, and t is the time of cultivation in the logarithmic phase.

The above differential equations were solved simultaneously using a fourth-order Runge–Kutta method (at that time — MATLAB was differently viewed) to arrive at some predictions of the expected DO profiles, even before the experiments were done. This gave us some confidence that the method would work before doing the experiments to demonstrate it. A sample from the predictions is presented in Figure 5.4, which shows that it was possible to maintain DO above 50% through the appropriate addition of H_2O_2 pulses.

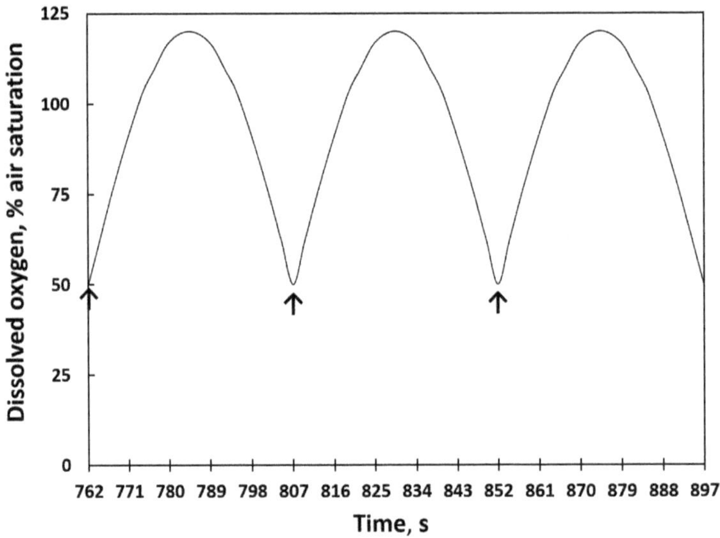

Figure 5.4. Dissolved oxygen profiles with LPOS simulation. The arrows represent H_2O_2 pulse additions

5.5. The Surprise with the Mechanism of Oxygen Availability Through LPOS

The initial guess regarding the mechanism of oxygen availability from H_2O_2 when LPOS is used was that the cell will somehow sense H_2O_2 in the surrounding space and secrete catalase to decompose it to liberate oxygen in the extracellular space. However, it occurred to me that H_2O_2 is a small, uncharged molecule that can easily pass through the cell membranes. If so, is the extracellularly added H_2O_2 getting decomposed in the intracellular space? To answer this question and gather insights into the mechanism of oxygen availability through LPOS, we did the studies described in this section. The details on the various aspects presented in this section are available in Sriram and Sureshkumar [2000].

5.5.1. *Where is the catalase?*

To find the location of H_2O_2 decomposition, we first conditioned the *X. campestris* cells through exposure to H_2O_2. We did separate experiments with the "conditioned cells" and the "medium without cells" — the medium in which cells were earlier present while being conditioned to H_2O_2. When the conditioned cells were exposed to H_2O_2, they generated oxygen at a rate of 6.77 mM h^{-1}. However, when the medium without cells was exposed to the same concentration of H_2O_2, the oxygen generation rate was only 0.5% of the rate with conditioned cells. That oxygen generation rate was comparable to the rate of oxygen dissolution across the air–liquid interface. This suggests that the oxygen dissolution from the air was solely responsible for the oxygen increase in the medium when cells were not present. More importantly, no catalase was present in the medium that earlier contained the cells when exposed to H_2O_2. Therefore, we ruled out the possibility of cells sensing H_2O_2 and secreting catalase to decompose it, with some further arguments on the impossibility of catalase degradation in the extracellular space during the experiment. Also, based on NADH fluorescence measurements, the basis of which is described in Sriram and Sureshkumar [2000], we ruled out the possibility of cytoplasmic decomposition of the H_2O_2 that enters the cells, at least until 1.6 mmole $(g\ cell)^{-1}\ H_2O_2$. Thus, we deduced that the decomposition of the H_2O_2 that enters the cells happens in the periplasmic space (the space between the inner and outer membranes of the bacterial cell) until 1.6 mmole $(g\ cell)^{-1}\ H_2O_2$. To avoid toxicity, it would be good to operate below that H_2O_2 level. Further, the oxygen uptake rates when LPOS is used compared well with those when aeration is used. In other words, LPOS does not alter the oxygen uptake rates of the cell.

5.5.2. *Excess oxygen, most likely, is ejected by the cell*

Cells are known to maintain their cytoplasmic oxygen levels at low values of around 30 µM [Koch, 2002]. However, the amount of oxygen generated by catalytic decomposition of added H_2O_2 pulse(s) would be much higher than that value. Thus, it is reasonable to expect that the excess oxygen generated in the cytoplasm is pushed out into the extracellular space by the cell. Recall that the DO level increases almost immediately after H_2O_2 pulse addition. That DO increase seems to result from the oxygen pushed out by the cell. Thus, the whole concept of DO needs to undergo a significant change when LPOS is used for cultivation. The LPOS seems to completely fulfil the cell's oxygen needs at the place of need (intracellular), and the excess seems to be pushed out into the medium to increase the DO. However, the actual mechanism by which oxygen is ejected out of the cell, particularly whether an active component is involved, still needs to be understood.

5.5.3. *The flux of H_2O_2 seems to be energetically controlled by the cell*

Based on our careful observations, we postulated that the cell controls the flux of H_2O_2 across its membrane, and the proton motive force (PMF) provides the energy needed for that control. We built a mathematical model based on the above ideas/principles (details in Sriram and Sureshkumar, [2000]), which could explain our results well. Then, experimentally, we proved the involvement of the PMF in H_2O_2 flux control. When the PMF was disrupted with an ionophore, carbonyl cyanide m-chloro phenylhydrazone (CCCP), the cell's control over the H_2O_2 flux was lost, which led to higher DO values that were in stoichiometric agreement with the added H_2O_2.

5.6. Other Cultivations with LPOS

We have demonstrated the successful application of LPOS to many cultivations, in addition to the *X. campestris* cultivation described in detail earlier in this chapter, through cultivations of about 10 different organisms, including bacteria, yeast, and molds with LPOS. We have better disseminated the salient results with the mold, *Aspergillus niger*, through a publication [Rawool *et al.*, 2001].

A. niger is a versatile mold with significant industrial importance. It is used to produce a variety of products, including enzymes such as catalase, lipase, protease, α-amylase, and naringinase. Under appropriate conditions, it is also a system of interest for producing antibiotics or organic acids such as acetic, citric, and other acids. It exists in two forms — pellet and filamentous forms. It was a challenge to supply oxygen at adequate rates in a bioreactor to both forms. When *A. niger* exists in a filamentous form, the challenge is significantly higher because the filaments (hyphae) are shear sensitive — the biological structure gets torn apart by shear stress in the fluid environment. Thus, it is challenging to employ high agitation or aeration rates, which can improve the k_La and, hence, the oxygen supply rate. On the other hand, when it exists in a pellet form, the diffusion of oxygen to the interiors of the, say, spherical pellet, at sufficient rates is a challenge.

We considered the pellet form, which was relevant to our experimental studies then. We developed a mathematical model to understand the oxygen profiles inside the *A. niger* pellet, and the extent of oxygen deprivation. The mathematical model was based on balances of oxygen and hydrogen peroxide, considering the pellet as the system to get their respective concentration profiles inside the pellet.

When aeration is used to supply oxygen, diffusion of oxygen occurs inside the pellet along with its consumption by the cellular parts inside the pellet. The fundamental mass balance on a species, *i*, with diffusivity

D_i in the spherical system (see Suraishkumar, [2014] for the details) yields

$$\frac{\partial c_i}{\partial t} + \left(v_r \frac{\partial c_i}{\partial r} + v_\theta \frac{1}{r} \frac{\partial c_i}{\partial \theta} + v_\emptyset \frac{1}{r\sin\theta} \frac{\partial c_i}{\partial \emptyset} \right)$$
$$- D_i \left(\frac{1}{r^2} \frac{\partial}{\partial r}\left(r^2 \frac{\partial c_i}{\partial r} \right) + \frac{1}{r^2 \sin\theta} \frac{\partial}{\partial \theta}\left(\sin\theta \frac{\partial c_i}{\partial \theta} \right) + \frac{1}{r^2 \sin^2\theta} \frac{\partial^2 c_i}{\partial \emptyset^2} \right) = R_i$$

$$\text{Eq. 5.6 — 1}$$

Cancellation of the irrelevant terms (see Suraishkumar, [2014] for the details), invoking the steady-state condition, and simplification results in

$$\frac{D_{O_2}}{r^2} \frac{1}{\rho} \frac{d}{dr}\left(r^2 \frac{dC_{O_2}}{dr} \right) = [OUR] \qquad \text{Eq. 5.6 — 2}$$

$[OUR]$ is the oxygen uptake rate. If $[OUR]$ is taken as a constant, then the differential equation can be solved with the following boundary conditions.

$$at\, r = 0, \quad \frac{dC_{O_2}}{dt} = 0 \text{ (symmetry boundary condition)} \qquad \text{Eq. 5.6 — 3}$$

$$at\, r = R, \quad C_{O_2} = C_{O_2}^R \qquad \text{Eq. 5.6 — 4}$$

The oxygen profile in the pellet, from the above, is

$$C_{O_2} = C_{O_2}^R - \frac{[OUR]R^2 \rho}{6D_{O_2}}\left(1 - \frac{r^2}{R^2} \right) \qquad \text{Eq. 5.6 — 5}$$

When LPOS is used, the oxygen inside the pellet is generated by the decomposition of H_2O_2 that diffuses into the pellet. The mass balance (Eq. 5.6 — 1) on H_2O_2 in this situation, with C_H being the H_2O_2 concentration, D_H being the diffusivity of H_2O_2, and k being the first order H_2O_2 decomposition rate constant, leads to

$$\frac{D_H}{r^2} \frac{d}{dr}\left(r^2 \frac{dC_H}{dr} \right) = kC_H \qquad \text{Eq. 5.6 — 6}$$

The boundary conditions:

$$at\ r = 0,\ \frac{dC_H}{dt} = 0 \text{ (symmetry boundary condition)} \qquad \text{Eq. 5.6}-7$$

$$at\ r = R,\ C_H = C_H^R \qquad \text{Eq. 5.6}-8$$

An analytical solution for the H_2O_2 profile inside the pellet was obtained using a variable change, $Z = C_H R$, and the consequent transformations.

$$C_H = \frac{C_H^R}{e^{\sqrt{\frac{k}{D_H R}}} - e^{-\sqrt{\frac{k}{D_H R}}}} \left(\frac{R}{r}\right)\left(e^{\sqrt{\frac{k}{D_H r}}} - e^{-\sqrt{\frac{k}{D_H r}}}\right) \qquad \text{Eq. 5.6}-9$$

The mass balance (Eq. 5.6 — 1) on oxygen in this situation leads to

$$\frac{1}{\rho}\left(\frac{D_{O_2}}{r^2}\frac{d}{dr}\left(r^2\frac{dC_{O_2}}{dr}\right) + R_{O_2}\right) = 0 \qquad \text{Eq. 5.6}-10$$

where R_{O_2} is the net oxygen generation rate, that is, the algebraic sum of the generation rate from H_2O_2 and the consumption rate by the cells in the pellet.

$$R_{O_2} = \frac{k}{2}\left\{\frac{C_H^R}{e^{\sqrt{\frac{k}{D_H R}}} - e^{-\sqrt{\frac{k}{D_H R}}}}\left(\frac{R}{r}\right)\left(e^{\sqrt{\frac{k}{D_H r}}} - e^{-\sqrt{\frac{k}{D_H r}}}\right)\right\} - [OUR] \quad \text{Eq. 5.6}-11$$

The boundary conditions for Eq. 5.6 — 10 are

$$at\ r = 0,\ \frac{dC_{O_2}}{dt} = 0 \text{ (symmetry boundary condition)} \qquad \text{Eq. 5.6}-12$$

$$at\ r = R,\ C_{O_2} = 0 \text{ (oxygen concentration at the surface is zero) Eq. 5.6}-13$$

This differential equation called for a numerical solution.

The profiles of oxygen inside the 2 mm diameter pellet when only aeration is used and when LPOS is used for oxygen supply are given in Figure 5.5. The Figure shows that when aeration is used, the oxygen

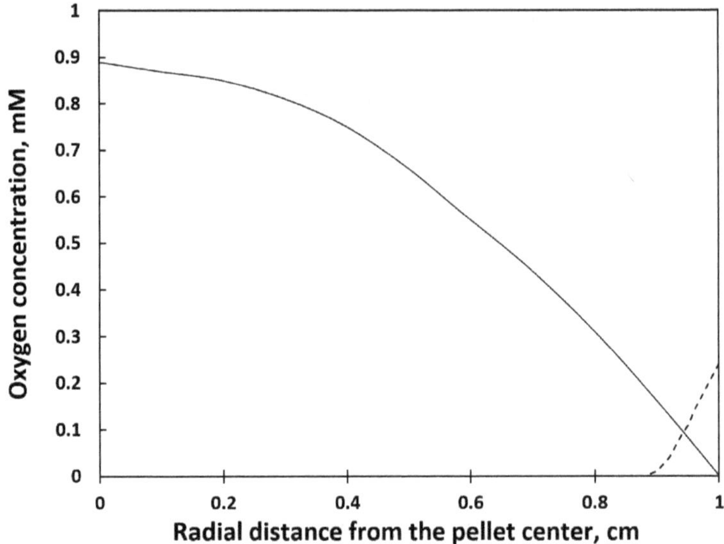

Figure 5.5. Simulated oxygen concentration profiles inside an *Aspergillus niger* pellet of 1 cm radius from a conventional cultivation (dashed line) and a LPOS-based cultivation (solid line)

concentration drops to zero at around 0.88 mm; thus, most parts of the pellet are devoid of oxygen. Whereas, when LPOS is used, the entire pellet has oxygen, with its concentration being highest at the pellet center. Thus, the oxygen supply challenge is entirely overcome by LPOS in the case of pellet morphology, too.

The growth rates, as inferred from the relevant constant in the logistic equation that we used to represent the growth mathematically, were comparable between aeration (0.15 h^{-1}) and LPOS-based (0.17 h^{-1}) cultivations. In addition, the specific level (amount produced per cell) of enzyme produced by *A. niger* was 71% higher for catalase and 56% higher for protease in the LPOS case compared to the aeration case. In a later chapter, we will consider other reasons for the increased enzyme productivity, apart from better oxygen supply.

Similarly, we have shown improvements in many cultivations with LPOS. It may be good to recall that photosynthetic organisms such as microalgae split water to generate oxygen during photosynthesis.

Thus, they may not need oxygen supply in ways similar to the non-photosynthetic organisms. However, the RS aspects to improve productivity are relevant even in photosynthetic systems, as we will see in a later chapter.

5.7. Optimization

The LPOS-based cultivations described in the previous sections in this chapter were usually carried out manually in response to the DO value. When the DO value reached a particular low value, the next pulse of H_2O_2 was added to the bioreactor. Thus, the LPOS-based cultivations were carried out in a nonoptimal fashion, with minimal understanding of the system.

When the understanding improved, we decided to do optimization studies to explore improvements in product formation. In the work by Chaitali *et al.* [2003], we did optimization studies on a fed-batch bioreactor for xanthan gum production. We built a process model based on material balances and posed it as an optimization problem. The objective was to maximize the volumetric concentration of xanthan gum at a particular time in the stationary phase of the cultivation. The solution to the optimization problem gave us profiles of H_2O_2, glucose, and a nitrogen source to achieve the objective.

We conducted experiments with the optimal feed profiles and achieved about a 1.5-fold increase in xanthan gum concentration compared to conventional cultivation. The optimal xanthan gum concentration was also about 50% higher than the non-optimal LPOS-based cultivations in which the H_2O_2 pulses were manually added. The specific details are available in the publication [Chaitali *et al.*, 2003].

5.8. How Expensive is LPOS?

To the uninitiated, the LPOS could appear expensive because it involves H_2O_2 solution. However, a closer look reveals a diametrically opposite fact that the LPOS is significantly less expensive than aeration. Let us look at some details here.

The reaction of interest

$$2H_2O_2 \xrightarrow{\text{catalase in cells}} 2H_2O + O_2$$

says that two moles of H_2O_2 produce one mole of oxygen. The molecular mass of H_2O_2 is 34, and that of molecular oxygen, O_2, is 32. The stoichiometry says that $34 \times 2 = 68$ g of H_2O_2 is needed to yield 32 g of O_2 upon complete breakdown.

Let us take the cost of

$$1 \text{ kg of } O_2 \text{ from } H_2O_2 = COST_{from\,H_2O_2}$$
$$1 \text{ kg of } O_2 \text{ from aeration} = COST_{from\,aeration}$$

These two costs cannot be directly compared because the ***effective*** cost of a kilogram of O_2 from H_2O_2 ($COST_{from\,H_2O_2,eff}$) is much less than that indicated by chemical stoichiometry. That is because

- The productivity is higher with LPOS (say, an average of 2.5-fold based on our collective experiences).
- The power input required to maintain a high k_La, which is required while usingair sparging to provide oxygen is about 20 times the power input required for reasonable mixing.
- When aeration is used, only a small fraction of the oxygen entering the bioreactor is utilized ($3/21 = 0.143$ or 14.3%), and the rest is lost.
- The sterilization and handling of the air supply system are nontrivial and leads to the loss of whole batches due to contamination (say, 10% of the batches are lost).
- No antifoam costs are involved while using LPOS.
- There are no major scale-up issues while using LPOS.
- The regulations (say, US FDA) on residual H_2O_2 are not a problem because there is no residual H_2O_2 in the medium.

We calculated the costs in the year 2000, with only the first four points considered. The ratio was surprising.

$$\frac{COST_{from\,H_2O_2,eff}}{COST_{from\,aeration}} = \frac{15\,paise}{200\,paise} = 7.5 \times 10^{-2}$$

Since then, the costs have gone up. The cost of a 50% solution of H_2O_2 when purchased in bulk was Rs. 14 per kg in 2000, and it is now (2021) Rs. 35 per kg. In the extreme, even assuming that the costs of oxygen from the air (aeration/filtration, etc.) have remained the same, the ratio is still a low 1.9×10^{-1}.

5.9. Can the LPOS Perform Better? — The Need to Handle RS

The LPOS addressed and solved the challenge of oxygen supply to cells in a bioreactor in a fundamental fashion. However, as with everything in science, things can be even better if the inherent issues with LPOS are addressed. A significant challenge is the toxicity due to RS generated when H_2O_2 interacts with Fe, Cu, and so on, that are present both inside and outside the cell. As we have seen in the earlier chapters, these RS could be beneficial or toxic depending on their specific (per cell) level. In the LPOS-based *X. campestris* cultivations, the maximum cell concentrations were only about 50% of those with aeration; the viable cell concentrations in the stationary phase were a much smaller percentage of the values with aeration. Thus, this was a challenge to be addressed to improve the bioreactor productivity even more. Our desire to address this challenge provided the entry for our research group sometime in 1995 into the vast but under-explored field of quantitatively manipulating RS in biological systems.

Sarkar *et al.* [2008] hypothesized that high concentrations of H_2O_2 exist in the feed zone for at least a certain time after the addition of a concentrated solution of H_2O_2, which led to local cell death in the feed zone and consequent reductions in maximum cell concentrations and bioreactor productivity. The H_2O_2 feed solution needs to be reasonably concentrated to avoid unacceptable increases in the bioreactor volume over the entire cultivation.

Through experiments and mathematical modeling, Sarkar *et al.* [2008] found that the feed-zone H_2O_2 concentrations were 12- to 14-fold higher than the bioreactor bulk H_2O_2 concentrations. Further, they explored various H_2O_2 feed strategies to minimize the toxic effects of H_2O_2.

When my first PhD student, Manjula Rao, joined our group in January 1996, she wanted to explore the idea of sensitizing the cells to higher H_2O_2 concentrations (she called it "vaccinating the cells") so that they become more robust. Thus, the toxic effects of H_2O_2 on bioreactor productivity can be reduced. The next chapter will begin presenting that interesting and far-reaching story.

References

Chaitali, M., Kapadi, M., Gudi, R. D., and Suraishkumar, G. K. (2003). Productivity improvement in xanthan gum fermentation using multiple substrate optimizations. *Biotechnol. Prog.*, 19, 1190.

Herbst, H., Schumpe, A., and Deckwer, W. D. (1992). Xanthan production in stirred tank fermenters. *Chem. Eng. Technol.*, 15, 425.

Hocker, H., Langer, G., and Werner, U. (1981). Mass transfer in aerated Newtonian and non-Newtonian liquids in stirred reactors. *Ger. Chem. Eng.*, 4, 54.

Koch, C. J. (2002). Measurement of absolute oxygen levels in cells and tissues using oxygen sensors and 2-nitroimidazole EF5. In: *Redox Cell Biology and Genetics, Part A*, edited by Sen C. K., and Packer L., Academic Press, New York.

Pinches, A., and Pallent, L. J. (1986). Rate and yield relationship in the production of xanthan gum by batch fermentation using complex and chemically defined growth media. *Biotechnol. Bioeng.*, 26, 1484.

Rawool, S., Sahoo, S., Rao, K. K., and Suraishkumar G. K. (2001). Improvement in enzyme productivities from mold cultivations using the liquid-phase oxygen supply strategy. *Biotechnol. Prog.*, 17, 832.

Sarkar, P., Ghosh, K., and Suraishkumar, G. K. (2008). High hydrogen peroxide concentration in the feed-zone affects bioreactor cell productivity with liquid phase oxygen supply strategy. *Bioproc. Biosys. Eng.*, 31, 357.

Sriram, G., ManjulaRao, Y., Suresh, A. K., and Sureshkumar. G. K. (1998). Oxygen supply without gas-liquid film resistance to *Xanthomonas campestris* cultivation. *Biotechnol. Bioeng.*, 59, 714.

Sriram, G., and Sureshkumar, G. K. (2000). The mechanism of oxygen availability from hydrogen peroxide to aerobic cultures of *Xanthomonas campestris*. *Biotechnol. Bioeng.*, 67, 487.

Suraishkumar, G. K. (2014). *Continuum Analysis of Biological Systems: Conserved Quantities, Forces and Fluxes*, Springer-Verlag, Heidelberg.

Chapter 6

Inoculum RS Induction to Improve Bioreactor Productivities

We saw a strategy for oxygen supply in Chapter 5, the liquid-phase oxygen-supply strategy (LPOS), which eliminated a fundamental challenge to adequate oxygen supply in aerobic bioreactors. We also saw that the LPOS could be improved and optimized and that it could have deleterious effects. The deleterious effects led to a maximum cell concentration of only about 55% in LPOS-based cultivation (1.6 g L^{-1}) compared to that in the conventional, aeration-based cultivation (2.9 g L^{-1}), even though the volumetric xanthan gum concentration was 70% higher in the LPOS-based cultivation. Further, the viability percentage in the early-stationary phase was much lower in the LPOS-based cultivation than in conventional cultivation.

We suspected that the deleterious effects of LPOS that reduced the maximum cell concentration were mediated by reactive species (RS), such as the hydroxyl and superoxide species. The RS were expected to be generated when H_2O_2 interacts with Fe or Cu ions in the medium or, more relevantly, in the intracellular space. Our idea was to find ways of reducing the deleterious effects of RS on cells in culture so that the LPOS could be even more successful. As mentioned in an earlier chapter, Manjula, the first student who worked with me for her Ph.D., had the idea to "vaccinate cells" against the deleterious effects of LPOS. In other words, the idea was to sensitize cells to hydroxyl radicals, which are the most reactive and, thus, most deleterious among the RS. It was expected that such sensitized cells, subsequently, would be better tolerant to the hydroxyl radicals and possibly other radicals generated when LPOS is employed.

To sensitize the cells to H_2O_2, we thought of exposing the *Xanthomonas campestris* inoculum to a hydroxyl radical generator in an appropriate fashion — Manjula devised a three-stage treatment procedure with HOCl as the treatment agent [ManjulaRao and Sureshkumar, 2001]. Then, it was to be used for the cultivation with LPOS. However, we needed first to characterize the effects of H_2O_2 treatment under standard conditions. Thus, we cultivated treated cells (treated inoculum) in a bioreactor under conventional aeration. The results from that cultivation, which we reproduced a significant number of times for confidence, were surprising.

6.1. Inoculum RS Induction and Xanthan Gum Levels

With aeration as the oxygen source, the maximum specific growth rates and the maximum cell concentrations of the untreated and HOCl-treated cells were comparable, which was encouraging. The surprise was with the xanthan gum concentrations achieved. The 50-h xanthan gum concentration with the untreated cells was 3.1 g L^{-1}, whereas with HOCl-treated cells, it was 6.5 g L^{-1}, which was more than twofold higher. A mere treatment with a hydroxyl radical inducer, HOCl, more than doubled the volumetric product level. Further, the quality of the gum, as evidenced by the viscosity of its solution in water (0.2 g L^{-1} measured at 8 s^{-1}), was a significant 20% higher. Thus, HOCl treatment improved both the quality and quantity of the product, xanthan gum. In addition, we realized that the earlier increase in xanthan gum concentrations with LPOS that we saw in Chapter 5 could have resulted from better aeration and induced RS effects.

Further, we investigated the reason for the increase in gum quality. Xanthan gum is a polymer of glucose residues with side chains. The side chains contain acetate and pyruvate. The extent of acetylation inversely correlates with gum quality, whereas the extent of pyruvylation directly correlates with gum quality [Hassler and Doherty, 1990]. Table 6.1 presents the mass percentages of acetate and pyruvate in the two cultivations. The data show that, as expected, the pyruvate mass percentage in

- ROS activate the *SoxRS* regulon (SoxR and Sox S proteins), whose levels remain high in subsequent generations as an adaptive response
- SoxS binds to the strong promoter of the *gum B* gene

- This leads to higher mRNA production, which further leads to higher gum production

Figure 6.1.   Our hypothesis for the increase in xanthan gum production by *Xanthomonas campestris* in response to inoculum RS induction

Table 6.1.   The mass percentages of pyruvate and acetate in xanthan gum produced by untreated and HOCl-treated cells of *Xanthomonas campestris*

	Untreated cells	**HOCl-treated cells**
Pyruvate, mass %	1.9	3.1
Acetate, mass %	5.7	3.3

HOCl-treated cells was 63% higher, and the acetate mass percentage was 42% lower than in untreated cells.

This discovery that a mere treatment of the inoculum with HOCl can significantly improve the quality and quantity of the product was far-reaching and exciting. Our core research took an interesting turn into RS and its manipulations in biological systems toward various objectives, which would keep us occupied for the next 25+ years and still continue.

Naturally, we were interested in finding the mechanism of increase in the product yields with HOCl treatment. Under normal conditions (no excess RS), the expressions of genes relevant to the xanthan gum production through metabolic pathways were known to be regulated by a strong promoter [Katzen *et al.*, 1996]. That promoter was located upstream of the *gum B* gene. We hypothesized that the HOCl-induced RS activated the same promoter and thus ultimately increased the xanthan gum production.

The details of our hypothesis (Figure 6.1) for the genetic basis for the improvement were as follows [ManjulaRao and Sureshkumar, 2001]: RS activate the *SoxRS* regulon that consists of the SoxR and SoxS transcription factors. The levels of SoxR and SoxS proteins remain high in subsequent generations as an adaptive response. SoxS binds to the strong promoter of the *gum B* gene. This leads to higher mRNA production, which further leads to higher gum production. We experimentally proved the SoxS–DNA binding and transcriptional activation. We identified the region where SoxS binds through inspection of the *gum B* promoter region and showed through EPR spectroscopy that SoxS binds explicitly to that region. The mRNA concentration obtained in the presence of SoxS was 2.4-fold higher than the value without SoxS.

Do you recall our reason for treating the cells comprising the inoculum with HOCl in the first place? The HOCl treatment was expected to sensitize the cells to RS so that they would better tolerate the RS generated by H_2O_2 pulses used for LPOS. We did experiments with HOCl-treated cells grown in cultures in which LPOS was employed to provide oxygen as a part of the studies designed to better understand the RS and antioxidant effects on the growth and xanthan gum productivity of *X. campestris* as summarized next.

6.2. Insights Into the Relationships Between RS Levels and Cultivation Outcomes

To understand the effects of RS on *X. campestris* growth and its xanthan gum productivity, we did experiments with untreated and HOCl-treated cells grown in the presence and absence of antioxidants, namely, ascorbic acid, tocopherol, and a mixture of the two. Oxygen was supplied through the LPOS. In addition, we developed a mathematical model based on relevant RS reactions. The details of the summary presented next in this section are available in the relevant publication [ManjulaRao *et al.*, 2003].

Let us first look at the mathematical model. We considered some relevant and prominent RS initiation and RS termination reactions, as shown in Figure 6.2. The RS of interest were hydroxyl, superoxide, and peroxyl radicals. As shown, the reactions included some prominent ones with antioxidants (when present), too. In addition, the effects of the ratio of Fe^{2+} to Fe^{3+} ion concentrations on the radical pool and the expected effects

Initiation

$$H_2O_2 + Fe^{2+} \longrightarrow Fe^{3+} + OH^- + {}^{\bullet}OH$$

$$H_2O_2 + Cu^+ \longrightarrow Cu^{2+} + OH^- + {}^{\bullet}OH$$

$$O_2 + Fe^{2+} \longrightarrow Fe^{3+} + O_2^{\bullet -}$$

$$ROOH + Fe^{3+} \longrightarrow Fe^{2+} + H^+ + ROO^{\bullet}$$

Termination

$${}^{\bullet}OH + {}^{\bullet}OH \longrightarrow H_2O_2$$

$$O_2^{\bullet -} + O_2^{\bullet -} + 2H^+ \xrightarrow{SOD} H_2O_2 + O_2$$

$$HOO^{\bullet} + HOO^{\bullet} \longrightarrow H_2O_2 + O_2$$

$$HOO^{\bullet} + O_2^{\bullet -} + H^+ \longrightarrow H_2O_2 + O_2$$

$${}^{\bullet}OH + HOO^{\bullet} \longrightarrow H_2O + O_2$$

$${}^{\bullet}OH + O_2^{\bullet -} \longrightarrow OH^- + O_2$$

With vitamins C and E

$$Vit\ C^- + O_2^{\bullet -} + H^+ \longrightarrow H_2O_2 + Vit\ C^{\bullet -}$$

$$Vit\ C^- + {}^{\bullet}OH \longrightarrow H_2O + Vit\ C^{\bullet -}$$

$$Vit\ E + O_2^{\bullet -} \longrightarrow HO_2^- + Vit\ E^{\bullet}$$

$$Vit\ E + ROO^{\bullet} \longrightarrow ROOH + Vit\ E^{\bullet}$$

Figure 6.2. The reactions that formed the basis for the mathematical model developed to understand RS effects

Table 6.2. Comparison between the $[Rad^{\bullet}]$ values obtained from simulations (numbers in plain text) and predictions (numbers in boldface) of the mathematical model and the experimentally obtained values. The initial concentration of vitamin C was 0.5 mM, and that of vitamin E was 0.1 mM

	Aeration	LPOS	LPOS + vit C	LPOS + vit E	LPOS + vit C + vit E
From the experiments (average value)	0.08	2.09	2.50	0.22	0.25
From the mathematical model	**0.08**	**2.09**	2.49	0.22	**0.26**

of vitamin C and vitamin E on specific growth rate were also represented in the mathematical model. The RS measurements were made through EPR spectroscopy, and thus, the predominant RS species were known. However, the analysis considered a combination of RS, represented as $[Rad^{\bullet}]$, where $[Rad^{\bullet}] = [{}^{\bullet}OH] + [O_2^{\bullet -}] + [ROO^{\bullet}]$ Taking into account the pseudo-steady-state conditions by which the initiation rate was equated to the termination rate, the simulated and predicted values of $[Rad^{\bullet}]$ are compared with the experimental values measured in the log phase (2.5 h) in Table 6.2. The values in boldface are the predicted values. It can be seen

Table 6.3. The growth rate constant, k, from the logistic equation under various LPOS cultivations of HOCl-treated cells. The initial concentration of vitamin C was 0.5 mM and that of vitamin E was 0.1 mM

	LPOS	**LPOS + vit C**	**LPOS + vit E**	**LPOS + vit C + vit E**
Growth rate constant, k, h^{-1}	0.15	0.27	0.95	0.90

from the table that the values from the mathematical model, including the predicted ones given in bold, are close to the experimental values.

The culture growth in the various cultivations was mathematically described well using a logistic equation.

$$\frac{dx}{dt} = k\left(1 - \frac{x}{x_S}\right)x \qquad\qquad \text{Eq. 6.1}$$

The values of the growth rate constant, k, for the different cultivations, are given in Table 6.3.

As seen from the table, in the presence of 5 mM vitamin C (initial concentration) alone, it increased to 0.27 h^{-1}. In the presence of 1 mM vitamin E (initial concentration), the growth rate constant was significantly higher at 0.95 h^{-1}. When both 5 mM vitamin C and 1 mM vitamin E were present, it decreased to 0.90 h^{-1}.

When the growth rate constants from the above and other relevant cultivations were plotted against the log-phase (2.5 h) specific intracellular ROS level in mmole (g cell)$^{-1}$, as expected, a linear negative correlation was obtained:

$$k = -0.39\ (intracellular\ ROS\ level) + 1.18 \qquad \text{Eq. 6.2}$$

A more exciting variation resulted when the specific xanthan gum levels, given in Table 6.4 for the LPOS cultivations of HOCl-treated cells, as well as the levels obtained from aeration experiments, were plotted against the corresponding stationary phase (50 h) specific intracellular RS levels. The trend is shown in Figure 6.3. In contrast to expectations, the specific xanthan gum level increased with the specific intracellular RS level until a particular value before it showed the expected decreasing

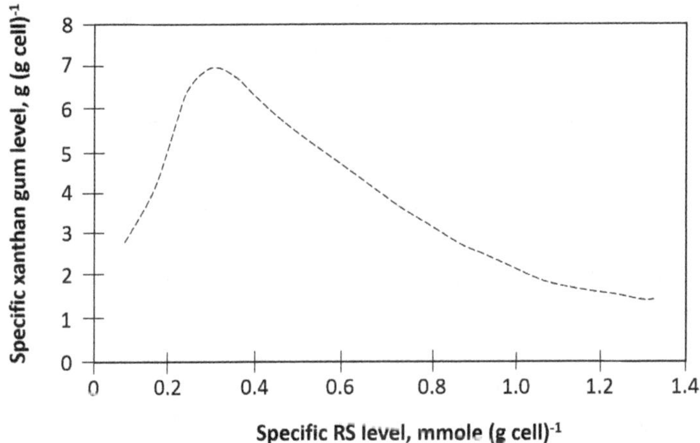

Figure 6.3. The relationship (trend) between specific xanthan gum level and specific intracellular RS level. There is an optimum level

Table 6.4. The specific xanthan gum levels under various LPOS cultivations of HOCl-treated cells. The initial concentration of vitamin C was 0.5 mM, and that of vitamin E was 0.1 mM

	LPOS	LPOS + vit C	LPOS + vit E	LPOS + vit C + vit E
Specific xanthan gum level (50 h, average value), $g \ (g \ cell)^{-1}$	7.00	2.25	6.40	4.70

trend. Thus, we realized that if we operate at certain specific intracellular RS levels, we can increase the xanthan gum productivity by each cell. There seems to be a particular range in which intracellular RS levels benefit bioreactor productivity.

6.3. Inoculum RS Induction in the *Aspergillus Niger* for Better Enzyme Productivity

After the significant success of the inoculum RS induction strategy in improving xanthan gum productivity from *X. campestris* cultivation, we wanted to explore whether inoculum RS induction can be used to enhance

Table 6.5.   Average values of the specific enzyme levels (U per g-cell) from *A. niger* cultivations with and without inoculum RS induction

	α-amylase	Protease	Glucose oxidase	Catalase
Untreated cells	284	0.11	160	669
Inoculum RS-induced cells	772	0.38	569	2827

the productivity of other biologicals, such as enzymes, which have a large annual turnover. The inoculum RS induction strategy has improved productivity in many systems that we have studied thus far. In the remaining part of this chapter, let us look at some of those systems.

Let us begin with enzyme production by the mold, *Aspergillus niger*, a widely used system in the industry. The details of this work are given in Sahoo *et al.* [2003]. A suitable inoculum RS induction treatment strategy was decided upon through appropriate preliminary experiments that considered the toxicity of the treatment agents. Let us consider the results from the experiments in which aeration was used as the source of oxygen to cultivate the HOCl-treated inocula.

The maximum cell concentration obtained in the inoculum RS induced *A. niger* cultivation was 86% higher than in the cultivation with untreated *A. niger*. Table 6.5 presents the comparison of specific enzyme levels for four of the enzymes produced by this system. From Table 6.5, it can be seen that the specific levels (U per g-cell) of α-amylase, protease, glucose oxidase, and catalase in cultivations with inoculum RS induced cells were 170%, 250%, 260%, and 320% higher, respectively, compared to the cultivations with untreated cells.

6.4. Inoculum RS Induction in the *Bacillus Subtilis* for Better Enzyme Productivity

Bacillus subtilis is an industrially relevant bacterium that is also well-studied. It is also a Generally Regarded As Safe (GRAS) organism, which increases its preference for industrial use. It is a versatile system that can produce several products depending on culture conditions; the products include enzymes, other proteins, and other molecules relevant in medicine and agriculture. With an appropriate inoculum RS induction procedure

Table 6.6. Average values of the specific enzyme levels (U per g-cell) in the extracellular space from *Bacillus subtilis* cultivations with and without inoculum RS induction

	Untreated inoculum	HOCl-treated inoculum
α-amylase, U (g cell)$^{-1}$	74.6	167.2
Protease, U (g-cell)$^{-1}$	0.22	0.56

(the induction procedure and further details of the aspects mentioned in this section are available in the paper [Mishra *et al.*, 2005]), a good improvement in the specific amounts of extracellular α-amylase and protease was achieved as given in Table 6.6. The oxygen supply for these cultivations was through aeration.

The *extracellular* specific enzyme levels are given in Table 6.6. We know that the enzymes are synthesized in the cell and are then secreted out. By now, we were reasonably clear that the specific intracellular enzyme levels were higher with inoculum RS induction, which we confirmed in this system, too [Mishra *et al.*, 2005]. However, Surabhi wanted to understand whether increased secretion rates could also contribute to the increased extracellular enzyme levels. At that time, in the early 2000s, microarray data for various organisms under various conditions were beginning to be freely available through data banks. Surabhi decided to analyze the microarray responses of *B. subtilis* to RS induction. The data in the databases then were for transcriptional responses to H_2O_2 exposure. We know that H_2O_2 is a milder inducer of RS compared to HOCl (proof given in Mishra *et al.*, [2005]). Thus, the available data would provide us with insights into the role of secretion in increased extracellular specific enzyme levels on inoculum RS induction. Through a detailed analysis of the microarray data, Surabhi found that the transcription of the *sipT* gene was increased 4.25-fold when *B. subtilis* was exposed to peroxide. The signal peptidase T (sip T) protein plays an important role in α-amylase secretion. Thus, it is possible that the increased secretion also contributed to the observed increase in specific extracellular levels of α-amylase.

In addition, we investigated [Mohan *et al.*, 2010] the membrane characteristics under RS-induced stress by studying its electrical properties. Please recall that the lipid bilayer cell membrane can be considered a capacitor because it has charged surfaces due to the charged heads of the amphipathic molecules separated by a dielectric material, the hydrophobic tails of the amphipathic molecules in the middle. The various electrical parameters, such as phase angle (between the voltage and current of the capacitor), impedance, capacitance, and breakdown voltage, provide insights into the packing of the lipid molecules in the bilayer and, hence, its fluidity. More the fluidity, more the leakage, and hence the higher rate of H_2O_2 into the cell as well as a higher rate of movement of intracellular proteins out of the cell across the cell membrane.

B. subtilis cells were exposed to three different extents of oxidative stress — no stress, mild stress (2.5 μM H_2O_2 added every 30 min), and strong stress (2.5 mM H_2O_2 + 100 μM $FeSO_4$ added every 30 min). The lipids from these cells were extracted, assembled into bilayer lipid membranes (BLMs), and their electrical and chemical properties were studied. We found that the oxidative stress modifies the electrical properties of the BLM. The decrease in phase angle, decrease in impedance with voltage and frequency, increase in capacitance with increasing AC voltage, and a reduction in breakdown voltage all show that the cell membrane under oxidative stress is not as ordered and, thus, more fluid than the cell membrane under no stress conditions.

A more fluid lipid bilayer could result from many molecular possibilities. An increase in the unsaturation of the fatty acid chains and lipid peroxidation are two among them. Our investigations [Mohan *et al.*, 2010] showed that the unsaturated lipid fraction increased by 127% and lipid peroxidation increased by 246% in cells exposed to strong stress compared to those exposed to no stress.

Thus, a combination of higher intracellular productivity and better secretion through a more fluid membrane can be used to improve enzyme productivity under oxidative stress conditions. In this study, the specific levels in the extracellular space of protease and α-amylase increased by about 62% and 137%, respectively, under oxidative stress induced by H_2O_2.

6.5. Inoculum-RS-Induction for Better Productivity of Other Products by *X. campestris*

Manjula discovered that *X. campestris* produces pyocyanin, an antimicrobial agent [ManjulaRao and Sureshkumar, 2000a]. The presence of pyocyanin was confirmed through IR spectroscopy and NMR spectroscopy. The specific pyocyanin level obtained from untreated cells was 4.2 mg (g cell)$^{-1}$. Interestingly, inoculum RS induction significantly increased pyocyanin production. Treatment with HOCl, menadione, and H_2O_2 increased the specific pyocyanin levels by 4.4-fold, 4-fold, and 2.6-fold, respectively. The HOCl-induced increase was dependent on the initial glucose concentration in the medium.

Manjula also found that *X. campestris* can produce high yields of ascorbic acid (vitamin C), a high-volume product, directly from glucose with inoculum RS induction [Manjula Rao and Sureshkumar, 2000b]. The direct production of ascorbic acid from glucose was known only in eukaryotes at low yields at that time.

Thus, inoculum RS induction seems to be a widely applicable strategy for improved product yields.

References

Hassler, R. A., and Doherty, D. H. (1990). Genetic engineering of polysaccharide structure: production of variants of xanthan gum in *Xanthomonas campestris*. *Biotechnol. Prog.*, 6, 182.

Katzen, F., Becker, A., Zorreguieta, A., Puhler, A., and Ielpi, L. (1996). Promoter analysis of the *Xanthomonas campestris* pv. campestris *gum* operon directing biosynthesis of the xanthan polysaccharide. *J. Bacteriol.*, 178, 4313.

ManjulaRao, Y., and Suraishkumar, G. K. (2000a). Oxidative-stress-induced production of pyocyanin by *Xanthomonas campestris* and its effect on the indicator target organism, *Escherichia coli. J. Ind. Microbiol. Biotechnol.*, 25, 266.

ManjulaRao, Y., and Sureshkumar, G. K. (2000b). Direct biosynthesis of ascorbic acid from glucose by *Xanthomonas campestris* through induced freeradicals. *Biotechnol. Lett.*, 22, 407.

ManjulaRao, Y., and Suraishkumar, G. K. (2001). Improvement in bioreactor productivities using free radicals: HOCl-Induced overproduction of xanthan gum from *Xanthomonas campestris* and its mechanism. *Biotechnol. Bioeng.*, 72, 62.

ManjulaRao, Y., Suresh, A. K., and Suraishkumar, G. K. (2003). Free radical aspects in *Xanthomonas campestris* cultivation with the liquid phase oxygen supply strategy. *Proc. Biochem.*, 38, 1301.

on cells, we need a specialized cultivation environment with defined flow conditions. The defined flow conditions would help us estimate or calculate the average shear stress (and shear rate) on the cells. From now onwards, we will use the terms "defined flow" and "defined shear" to denote this condition.

It was well established at that time (late 1990s) in the literature that at least two laminar flow environments provide defined shear. One is the Couette flow (named after Professor Maurice Couette, who contributed to that area) in the annular space between two concentric cylinders with either or both cylinders rotating. The other is the cone-and-plate geometry, with the angle between the cone and the plate being very small, say 0.5 to 1°. Since we wanted to carry out entire cultivations of about, say, 200 mL at least, under defined shear conditions, we chose the concentric cylinder geometry. The other geometry does not provide large volumes in reasonably sized cone-and-plate devices.

However, until then, the concentric cylinder geometry had been used only for short times with aerobic cells because it is difficult to provide

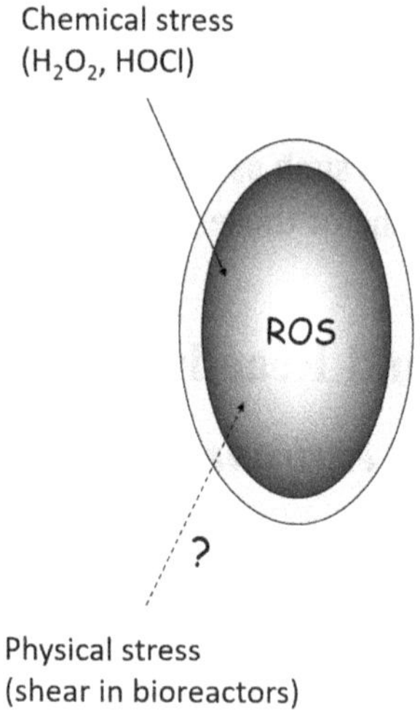

Figure 7.1. We had earlier seen that the effects of chemical stresses were mediated by reactive species (RS). We were interested in knowing whether RS were also involved in mediating the effects of physical stresses on cells. The RS involvement in mediating physical stresses was unclear at that time

oxygen through established means in that geometry. The gas–liquid transfer would need air to be bubbled through the cultivation space, and the bubbling would destroy the defined flow conditions. We needed to carry out entire cultivations under defined shear. Hence, we needed a way to supply oxygen throughout the cultivation of, say, at least 48 h, depending on the cultivation. Therefore, we ideated and designed a suitable means to provide oxygen that will not disturb the defined flow profile in the annular space between the concentric cylinders. We cut holes in the inner cylinder and covered the curved surface of the inner cylinder in the annular space with a thin Teflon sheet (Figure 7.2). Oxygen is highly permeable across Teflon, and hence, it can permeate from one side of the inner cylinder to its other side through the Teflon membrane covering the holes; thus, oxygen supply to the cells growing in the annular space is ensured.

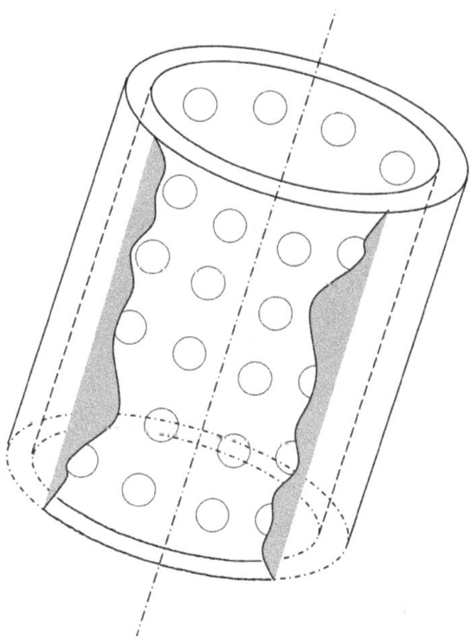

Figure 7.2. A schematic of the Couette flow device that shows the concentric cylinders and air holes on the inner cylinder that are covered with a Teflon membrane on the outer surface of the inner cylinder that faces the annular space. The outer cylinder rotates, while the inner cylinder is held stationary in this device. The cells are cultivated in the annular space between the two concentric cylinders

Table 7.1.   Measured or estimated values of the volumetric oxygen transfer coefficient, $k_L a$, in the Couette-flow bioreactor. The outer cylinder was rotated at the indicated rpm, while the inner cylinder was stationary

rpm	300	500	700	1000
$k_L a$, h^{-1}	4.54	4.56	4.58	4.59

The Teflon-membrane-attached smooth surface of the inner cylinder that faces the annular space poses no difficulties for the defined flow conditions to exist.

We did a large number of characterizations of the Couette-flow bioreactor (CFB), both experimental and mathematical-model based [Sahoo *et al.*, 2003], and measured or estimated the oxygen transfer capacity through the volumetric oxygen transfer coefficient, $k_L a$, at various rpm s of the outer cylinder. The values are shown in Table 7.1. Interestingly, the CFB $k_L a$ of about 4.5 h^{-1} compares well with 2.89 h^{-1}, the $k_L a$ in a shake flask subjected to 190 rpm. Thus, the CFB is well suited for studying the effect of shear alone on cells in cultivation without interference from oxygen transfer aspects.

The velocity profile, v_θ, in the annular space at any radial position, r, can be derived [Suraishkumar, 2014] to be

$$v_\theta = \frac{\Omega R\left(\dfrac{kR}{r} - \dfrac{r}{kR}\right)}{\left(k - \dfrac{1}{k}\right)}$$
Eq. 7.1

where R is the radius of the outer cylinder, k is the ratio of the outer to inner cylinder radii, and Ω is the angular velocity of the outer cylinder. The shear rate at any radial position can be derived [Sahoo *et al.*, 2003] by differentiation of the velocity profile to be

$$\dot{\gamma} = 2\Omega\frac{\left(\dfrac{kR}{r}\right)^2}{\left(k^2 - 1\right)}$$
Eq. 7.2

Table 7.2. Calculated shear rate values in the Couette-flow bioreactor. The outer cylinder was rotated at the indicated rpm, while the inner cylinder was stationary

rpm	300	500	700	1000
Shear rate, s^{-1}	445	741	1111	1482

An analysis of Eq. 7.1 for $k < 1$ shows that the tangential velocity, v_θ, varies almost linearly with the radius. Therefore, for flow in thin annuli where the stable laminar tangential flow can exist, the shear rate (the derivative of the tangential velocity with radius) at any r is a constant value. Thus, the shear rate experienced by the cells in the **thin** annular space can be approximated as a constant. All cells in the annular space experience the same shear rate. The shear rates at various rpm in the CFB are given in Table 7.2.

The shear rate in the shake flask shaken at 190 rpm was estimated (details in Sahoo *et al.* [2003]) to be 0.028 s^{-1}. Let us recall that the $k_L a$ in the shake flask was comparable to those in the CFB. Therefore, the low shear rate in the shake flask allows us to consider the shake flask cultivations as control experiments for studying shear effects on cells in cultivations.

7.1.2. *Effect of shear rate on cell size, growth, and enzyme production*

We initially expected that cells grown under shear would be more elongated than those grown under control conditions (low shear rate in the shake flask). However, the results surprised us. The cell sizes, as measured through scanning electron microscopy (SEM), showed that the average length of *Bacillus subtilis* cells was 3.1 μm at 0.028 s^{-1}, and it decreased by about 50% to around 1.5 μm at 1482 s^{-1}. The unexpected reduction could have resulted because the cells prefer to expose a smaller surface to the surroundings at high shear rate conditions.

The growth profiles at the highest and lowest shear rates employed were significantly different. The maximum cell concentration obtained in the control (shake flask, 0.028 s^{-1}) was 2.7 g L^{-1}, whereas at 1482 s^{-1}, it was 7.8 g L^{-1}, a 2.9-fold increase. The specific growth rate was also higher; at 1482 s^{-1}, it was 0.76 h^{-1}, about 3.5-fold higher than 0.22 h^{-1} at 0.028 s^{-1}.

The specific extracellular enzyme levels also increased significantly with the shear rate. The catalase specific level at 1482 s^{-1} was 3035 U (g-cell)$^{-1}$, about 7.7-fold of the value obtained at 0.028 s^{-1}. The protease specific level at 1482 s^{-1} was 0.73 U (g-cell)$^{-1}$, approximately 1.9-fold the value obtained in the control (0.028 s^{-1}). However, the α-amylase specific levels did not vary significantly with the shear rate.

7.1.3. *Reactive species levels at different shear rates*

The specific intracellular RS level (siROS, at 10 h — late log or early stationary phase) in cultivations done at various shear rates are presented in Figure 7.3. As can be seen from Figure 7.3, the 10-h siROS level increased with the shear rate, and the increase in the CFB cultivations (second bar onwards, from left to right) was linear with the shear rate.

7.1.4. *Mechanistic aspects of shear-induced reactive species*

The initial idea that physical stress, such as shear, could mediate its cellular effects through a molecular species, RS (Figure 7.1), was purely intuitive.

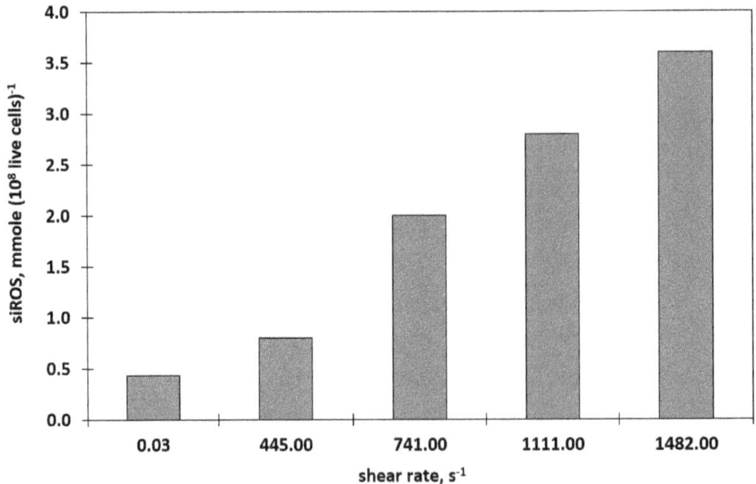

Figure 7.3. The relationship between specific intracellular reactive oxygen species level (siROS) and shear rate in *Bacillus subtilis*

When the observations in this work supported the idea, we were interested to derive insights into the mechanistic details. The question of how a physical quantity is transduced to a chemical (molecular) aspect in the cell was interesting. We suspected that NADH oxidase, an enzyme present on the cell envelope, which is physically impacted by shear, may be responsible; NADH oxidase is a known source of superoxide and superoxide-derived radicals. Through experiments with a NADH oxidase inhibitor, diphenyliodonium (DPI), we showed that NADH oxidase is the molecular transducer between the physical shear stress and the generation of molecular RS in the intracellular space.

It was known in the literature that the transcription factor σ^B was involved in the transcription of genes whose products offer protection against, say, oxidative, heat and salt stresses. The relationship between σ^B and shear stress was unknown at that time. Susmita was interested in exploring the relationship between shear stress and σ^B. She did the studies with a σ^B-mutant strain of *B. subtilis*, analyzed the data from different cultivations at different shear rates, and discovered an inverse relationship between specific σ^B levels and siROS levels in the cell, as depicted in Figure 7.4. Based on extensive studies with the σ^B-mutant, the details of

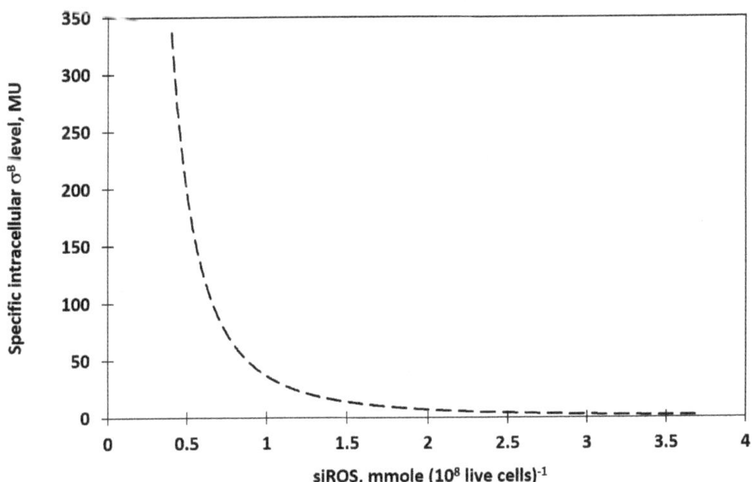

Figure 7.4. Relationship between specific intracellular sigma B level, expressed in Miller Units (MU), and specific intracellular reactive oxygen species level (siROS) in *Bacillus subtilis*

which can be found in Sahoo *et al.* [2003], we formulated the following mechanism for the mediation of shear effects on the cell: The cell senses the shear stress through the membrane-bound NADH oxidase, which generates RS — predominantly, superoxide and superoxide-derived radicals. The generated RS suppress the σ^B transcription and thus affect the protein levels under σ^B control. The outcomes of σ^B transcription suppression and decreased relevant protein levels are morphological changes, changes in growth, and enzyme production.

In the batch cultivations at 1482 s^{-1}, we noticed that the cell concentration dramatically reduced after reaching the maximum cell concentration at around 10 h. This dramatic decrease in cell concentration did not occur in cultivations at 0.028 s^{-1}. The cell concentrations were measured through cell scatter (OD); thus, we thought of two possibilities for the cell concentration decrease. One, the optical properties of the cell suspension changed, leading to a reduction in OD, or two, the viable cell concentration decreased dramatically after 10 h, when the cultivation was carried out at 1482 s^{-1}. The optical properties of the cell suspension can change with spore formation, and unusual cell death could dramatically decrease the viable cell concentration. We investigated both aspects and were richly rewarded. We discovered highly interesting and counterintuitive phenomena, as mentioned in the following two sections.

7.2. Shear Stress and Sporulation

As may be known, sporulation is the process of spore formation by the microorganism to survive unfavorable environmental conditions. The microorganisms develop a rigid cell envelope and transition to a non-vegetative state until the environmental conditions turn favorable. Sporulation can happen in nature as well as in batch bioreactors after the onset of the stationary phase. Thus, sporulation can lead to productivity loss in bioreactors, especially with secondary metabolites.

Sporulation is a cellular response to environmental stress. Thus, Susmita suspected that shear and the associated RS could impact sporulation, too — the relationship between shear stress and sporulation was unknown at that time.

Extensive work [Sahoo *et al.*, 2004] showed that counterintuitively, shear inhibits sporulation. The sporulation extent defined as

$$Sporulation\ extent = \frac{No\ of\ spores}{No\ of\ live\ cells\,(vegetative + spore)} \times 100$$

drastically reduced from 99% at 0.028 s^{-1} to about 1% even at 445 s^{-1}. We also showed that shear-induced RS regulate the inhibition of sporulation with the involvement of the general stress protein, Ctc, and general stress transcription factor, σ^B, which we saw earlier in this chapter.

7.3. Shear Stress and Programmed Cell Death in Bacteria

The investigation into the other reason for the dramatic decrease in cell concentration, specifically the viable cell concentration, led us to programmed cell death (PCD) in bacteria. Susmita wanted to study that aspect for her Ph.D. thesis, when I thought she had already done enough for her thesis. When Susmita suggested bacterial PCD, I was not sure. I was not sure because PCD in prokaryotes was controversial at that time in the literature. Her work until then was significant enough for a good Ph.D. thesis. Thus, the risk to her thesis from the work on PCD in bacteria was nonexistent, and I agreed.

This work [Sahoo *et al.*, 2006] showed for the first time that shear stress on bacteria can cause cell death, which resembles eukaryotic apoptosis. The activation of a caspase-3-(like)-protease (C3LP) followed by DNA fragmentation to cause a ladder pattern on the gel, which were the hallmarks of eukaryotic apoptosis, were found in *B. subtilis* cells subject to shear at 1482 s^{-1}. The work also showed that the shear-induced RS were directly responsible for the activation of C3LP as well as for DNA fragmentation.

7.4. Temperature and pH Stresses

We explored the relationship between pseudo-steady state RS levels and two other physical stresses, temperature and pH [Sarkar and Suraishkumar, 2011].

Temperature and pH are important bioreactor operation variables, too. At that time, the relationship between pH and RS was unknown, and no quantitative relationship was available between temperature and RS. We focused on superoxide in this study.

The superoxide levels measured through dihydroethidium (DHE) fluorescence at different temperatures and pH levels are presented in Figures 7.5 and 7.6, respectively. The specific intracellular superoxide levels increased with temperature and pH, respectively. We also studied many oxidative stress markers in the same work [Sarkar and Suraishkumar, 2011]. The markers included lipid peroxidation and some antioxidant levels. Some mechanistic aspects were also explored in that work.

In this work, the levels of antioxidant enzymes, namely superoxide dismutase and catalase, were measured. As pH changed from the physiological pH value (7) in either the acidic (pH < 7) or the basic (pH > 7) direction, the antioxidant enzyme levels increased. Whereas the RS (superoxide) level increased monotonically with pH between pH 5 and 9, the range studied. Thus, the general acceptance in the literature that the intracellular antioxidant enzyme levels are reliable biological markers of oxidative stress seems invalid. Although we did not pay attention to this

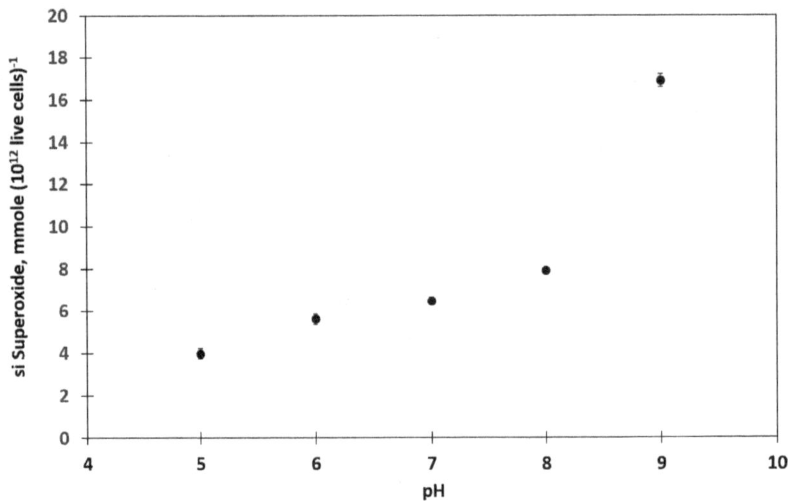

Figure 7.5. Variation of specific intracellular (si) superoxide levels with pH in *Bacillus subtilis*

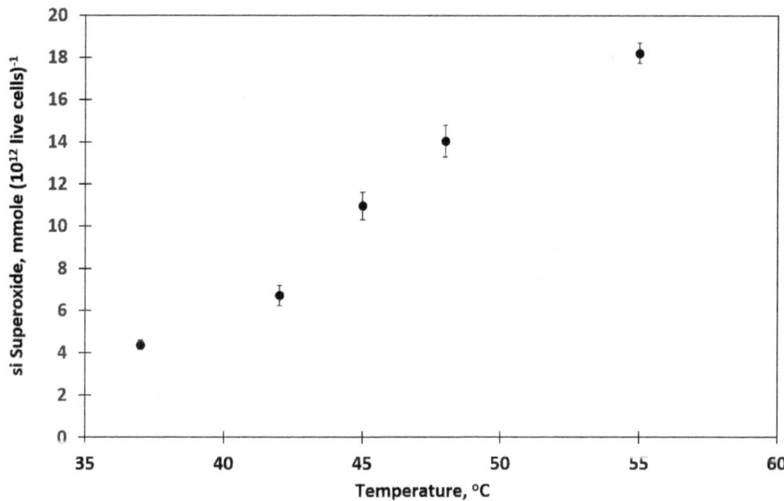

Figure 7.6. Variation of specific intracellular (si) superoxide levels with temperature in *Bacillus subtilis*

anomaly then, it turned out to be significant a few years later. We will see the significance of this anomaly in the chapter on cancer treatment and management (Chapter 9).

7.5. Light and Nutrient Stresses: Relationship Between Specific RS Levels and Specific Bio-oil Productivity

In the mid-2000s, there was a renewed interest worldwide in alternative fuels, especially biofuels, probably due to the increase in crude oil prices. It was a big interest wave, and I rode it, too, with the prodding and support of Shrikumar Suryanarayan, our generous alumnus. We got into biofuels from microalgae with the aim of producing affordable oil from the microorganism, preferably a marine species, as a replacement for crude oil from fossil sources deep in the earth. The thinking was that if they could be somehow grown successfully on the sea, with nutrients from the sea, we could economically produce bio-oil from marine microalgae. Many installations of crude oil are on high seas, and when their time runs out, the installations of microalgae farming on the sea to produce bio-oil could take their place. That way, the massive infrastructure for the distribution and use of crude can still be used for the microalgal bio-oil — only the source of the oil would be different.

Bio-oil is fundamentally neutral lipids, which get stored in microalgae. In principle, producing bio-oil from microalgae merely involves growing, harvesting, and squeezing them to extract oil. However, the main challenge is an economical way to do the above three processes, to compete and go below the costs of crude oil production. This challenge has, until today, precluded the possibility of widespread use of bio-oil from microalgae. Strategies to improve bio-oil yields would play a significant role in making it economically feasible. We decided to focus on this aspect through RS.

An analysis method that is typically used to extract lipids (bio-oil) for quantification or further analysis is the Bligh and Dyer method. During our studies, we realized that this method was also extracting chlorophyll in addition to lipids. Given the large amounts of chlorophyll in photosynthetic cells, the extent of overestimating lipids by the Bligh and Dyer methods could be as bad as 25%. Investigating deeper, we realized that the Bligh and Dyer method was originally developed many decades ago to extract lipids from animal cells that do not contain chlorophyll. Somewhere down the line, it was used to extract lipids from photosynthetic organisms that contain chlorophyll. We brought this to the attention of the researchers [Archanaa *et al.*, 2012] and also developed a method to extract lipids alone without chlorophyll contamination.

Let us first look at our work on quantifying the effect of specific RS levels on specific bio-oil productivity [Menon *et al.*, 2013]. This relationship arose from our investigations on the effect of light stress and nutrient stress on the growth and bio-oil production of the marine microalga *Chlorella vulgaris*. The effects of the stress caused by excess nutrients were studied by growing the organism in different media. All the components in the basal (f/2) medium, termed 1X medium, were linearly increased to 2x, 4x, 6x, 8x, 10x, 12x, and 16x, resulting in media that can provide different nutrient excess levels. The effects of light wavelength were studied through cultivations exposed to different LED lights; the light intensity at any wavelength was fixed at 4000 lux. Three wavelengths, 449 nm (blue),

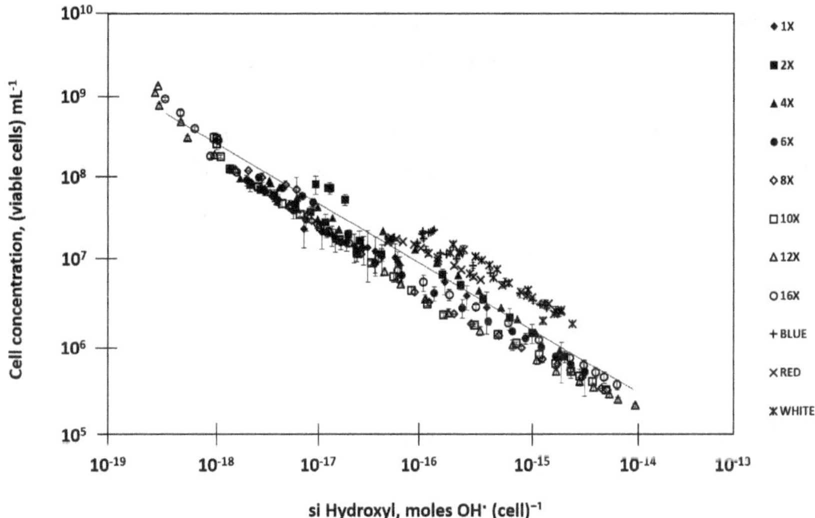

Figure 7.7.   The relationship between cell concentration and the specific intracellular (si) hydroxyl radical level, irrespective of the stress inducer employed in the study

654 nm (red), or 456 and 535 nm (white), were used, one for each cultivation. Significant results emerged from that study (only the most significant results are recreated here — more complete data are available in Menon *et al.* [2013]); let us look at some of those results in this section.

When the cell concentration and specific intracellular RS (hydroxyl) level data from all the above cultivations were plotted on a log–log scale (Figure 7.7), an almost linear relationship with a negative slope was evident. The best-fit inverse power–law relation was:

$$Cell\ conc \cdot (cells\ mL^{-1}) = 3 \times 10^{-5}\,(siRS\,(\{mole\ hydroxyl\ radicals\}\,cell^{-1})^{-0.7}$$

Eq. 7.3

Note that these data from all the phases of batch growth from all the cultivations under nutrient excess conditions as well as under different light wavelengths were used to plot Figure 7.7. Thus, the inverse power law seems valid for a wide range of conditions. Further, high nutrient conditions and high frequencies of light were perceived as being "stressful" to the cells in the literature at that time. Our work showed that those so-called "stressful" conditions led to the highest cell concentrations and maximum specific growth rates. Hence, we argued that the "stressful" nature of such conditions is debatable.

When we studied the relationship between specific lipid produced and siRS in all the experiments (nutrient excess and different light wavelengths), we got the result in Figure 7.8. The corresponding power law relationship (direct power law) was

$$si\ neutral\ lipid\ (\{\mu g\ triolein\}\ cell^{-1}) = 3.8 \times 10^{4}$$
$$(siRS\ (\{mole\ hydroxyl\ radicals\}\ cell^{-1})^{-0.7}$$

Eq. 7.4

The above positive correlation suggested the possibility of improving neutral lipid yields through RS manipulations. We will see some details of the RS manipulations in the following two sections.

More importantly, the relationships found in this work, especially those in Figures 7.7 and 7.8, strongly imply that *RS mediate stress (most likely, any stress — physical or chemical) effects on the cell.* Nowadays, this

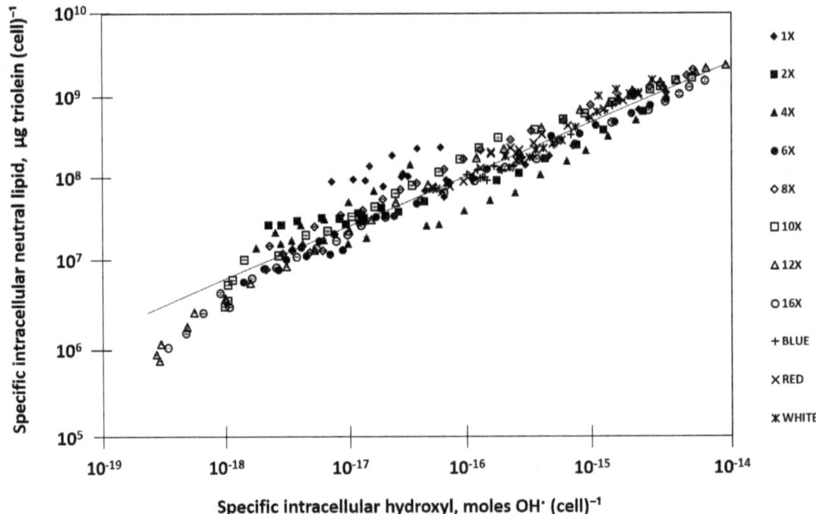

Figure 7.8. The relationship between the specific intracellular neutral lipid level (accumulated) and the specific intracellular hydroxyl radical level, irrespective of the stress inducer employed in the study

thought seems to be tacitly accepted. We seem to have grown up in this field in this period of acceptance, from a "maybe" to "tacit acceptance" over, say, 20+ years.

7.6. Reactive Species and UV-exposure-based Improvement in Bio-oil Production

As with many bioprocesses, the biological cell is the factory that synthesizes the product. Thus, in a given time for the process, it is best to have the highest possible number of individual factories and to ensure that each factory is operating at its highest capability to synthesize the product. In other words, having the highest specific growth rate and highest specific lipid content in the system is best. However, it is usually a challenge to achieve both the above goals simultaneously because the efforts to improve lipid content may interfere with the growth rate.

The work [Balan and Suraishkumar, 2014a], a part of which is discussed in this section, showed that it is possible simultaneously to increase both the specific growth rate and specific lipid content in the microalgal system, *C. vulgaris* through UV-induced RS manipulation. We investigated the effect of UVA (315–400 nm), UVB (280–315 nm), and UVA + UVB exposures during different stages of batch culture on the growth and lipid production in *C. vulgaris*. The UV exposure was for about 6 h every day, sandwiched between 5 h exposures to photosynthetically active radiation (PAR), and the day ended with exposure to 8 h of darkness (PAR (5 h) – UV (6 h) – PAR (5 h) – darkness (8h)). The control cultures were exposed daily to 16 h PAR – 8 h darkness.

The effects of UV exposure are summarized in Figure 7.9. It can be seen from Figure 7.9 that the specific growth rates were best increased (4.3-fold) by UVB exposure throughout the cultivation; the specific growth rate of the untreated cultivation was 0.132 h^{-1}. The 4.3-fold increase was comparable to UVB exposure in the early-log phase. Coincidentally, the maximum cell concentration was obtained with UVB exposure in the early-log phase. In comparison, UVA exposure in the early log phase increased the specific growth rate by about 3.2-fold.

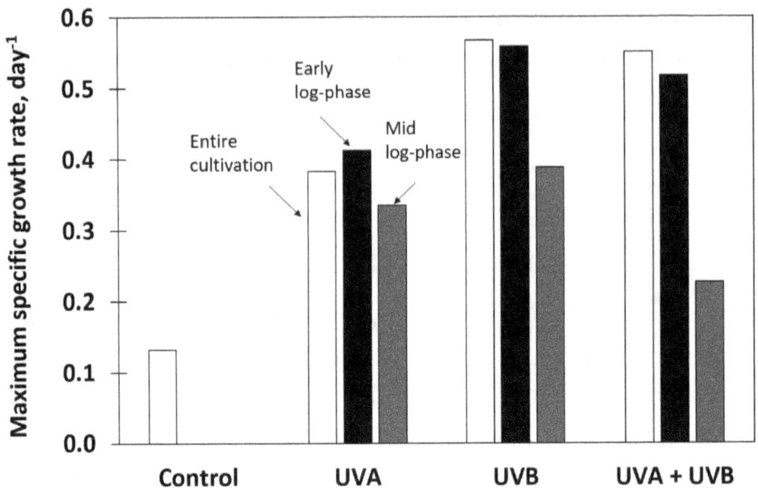

Figure 7.9.   Effect of different UV exposures on the maximum specific growth rate of *Chlorella vulgaris*

We investigated the mechanism of growth rate increase through hypotheses formulation and carrying out experiments to measure PSII activation, RS levels, antenna size, and others. The details are available in Balan and Suraishkumar [2014a]. The mechanism that emerged from those investigations was as follows: UVB excites PS II, which, in turn, increases the RS level, possibly through the excited triplet state of chlorophyll-a molecules. Increased RS level increases the antenna size, which is the number of chlorophyll molecules per unit PSII. The above lead to higher photosynthesis and, hence, better growth.

We also found that UVA treatment resulted in an 8.2-fold increase in lipid content compared to control. Among the lipids produced, saturated fatty acids (SFAs) are preferred over polyunsaturated fatty acids (PUFAs); thus, higher SFA:PUFA ratios are preferred. UVA treatment also resulted in a 3.2-fold increase in SFA:PUFA compared to control.

Based on the above findings that UVB increased specific growth rate, whereas UVA increased lipid content, we proposed and demonstrated a UVB/UVA treatment strategy to improve lipid productivity. Note that lipids are accumulated in the cell predominantly in the stationary phase after the growth phase is complete. Thus, the strategy was to use UVB treatment to maximize cell concentration, and then UVA treatment to improve lipid accumulation. The strategy worked well, and we achieved an 8.8-fold increase in lipid productivity (mg neutral lipid per day) compared to the control.

7.7. Reactive Species Rhythms

We discovered [Balan and Suraishkumar, 2014b] that the pseudo-steady-state (PSS) specific levels of RS (amount per cell) vary rhythmically with time. Please see Chapter 1 to understand the concept of PSS. An example of the rhythmic variation is given in Figure 7.10A. To the regular eye, multiple rhythms, say ultradian rhythms, seem to exist, especially when viewed in the time domain (time on the x-axis). Ultradian rhythms have a period of less than 24 h, whereas circadian rhythms have a period of about 24 h. Therefore, other analyses in the *frequency domain* need to be performed on the data to extract the rhythm characteristics. In this work, we did power spectral analysis using the Lomb–Scargle periodogram (LSP) to quantify the rhythm parameters and cosinor analysis to ensure the confidence level in our data. An example of an LSP is given in Figure 7.10B. The details of the analyses with references are available in Balan and Suraishkumar [2014b].

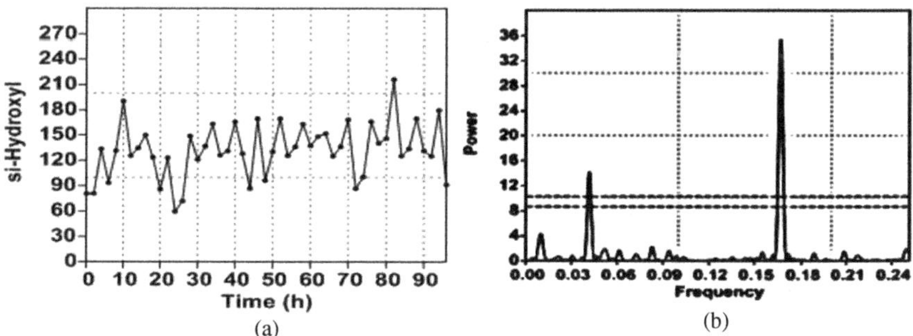

Figure 7.10. (A) Time profile of specific intracellular hydroxyl radical level (B) Frequency domain analysis of power to identify the dominant rhythm frequencies

7.7.1. *Reactive species rhythm reset to improve bio-oil productivity*

We found the rhythms in PSS levels of hydroxyl radicals to be endogenously (naturally) present in the cell even without any treatment. The cultivation was carried out with PAR (16 h) – dark (8 h) daily light scheme. Analysis of the power spectrum of the PSS levels showed the peak at a frequency of 0.167 h^{-1}, which translates to a time period of 5.99 h, an ultradian rhythm.

The photosynthetic machinery in cells produces RS, which are countered by several antioxidant molecules such as enzymes, pigments, and others; the cell adjusts the rates of the two opposing processes to maintain its RS homeostasis (redox homeostasis). When some process disrupts the RS homeostasis by generating RS at higher rates, the cellular processes sensitive to redox level would be activated/suppressed. Lipid formation is sensitive to the redox level or the RS level, as we have already seen in the previous sections.

Rhythm entrainment refers to rhythm reset, which synchronizes the intracellular oscillator to an external cue, such as a change in light properties. Rhythm entrainment would change the time-period/frequency of the existing rhythm. This adaptation to the external cue happens in all cells in the culture. Thus, all cells in the culture can be synchronized toward, say, lipid formation, which would lead to better lipid productivity. Further, entrainment to shorter period rhythms can be expected to cause faster switching between the possible extremes of the intracellular redox state. The faster switching between the possible extremes of the redox state is expected to benefit higher lipid formation. This is because a reduced redox state activates an essential enzyme for lipid synthesis, acetyl CoA carboxylase (ACCase) [Geigenberger *et al.*, 2005; Deitz and Pfannschmidt, 2011], whereas, subsequent oxidized redox state promotes the synthesis of neutral lipids such as triacylglycerols [Guarnieri *et al.*, 2011; Rismani-Yazdi *et al.*, 2012]. With the above ideas, we studied the effects of UV-based rhythm entrainment on lipid production in *C. vulgaris* cultures.

We changed the UV exposure pattern to the cultures to effect the entrainment. The modified daily light pattern was PAR (5 h) – UVA (6 h) – PAR (5 h) – dark (8 h). After the entrainment, the time period of the altered ultradian oscillation in the hydroxyl radical level was 2.97 h (a frequency of 0.337 h^{-1}) compared to a time period of 5.99 h for the control. Interestingly, a reduction in the time period increased the temporal window in which the neutral lipids accumulate every day, in the stationary phase of the culture (Figure 7.11). The data are given in a Cartesian plot here but in the form of rose plots in the paper. The data show that the neutral lipid accumulation window increased from 4.5 h during Zeitgeber time (ZT) 6–10.5 h to more than twice that window, 10.5 h, during ZT 7.5–18 h. Zeitgeber time refers to the time every day after the light is switched on for illumination. The Zeitgeber time varies from 0 to 24 h, with the time of lights-on being assigned the zero value. The specific amounts of neutral lipid (average accumulation over eight days in the stationary phase) also significantly increased by close to twofold in the interval of maximum neutral lipid accumulation in the rhythm reset condition. The SFA, which improves the oxidative stability of the bio-oil, also increased from 54% in control to 80% in the rhythm reset condition. Thus, hydroxyl rhythm entrainment/reset is a good strategy to improve the quantity and quality of bio-oil from microalgae.

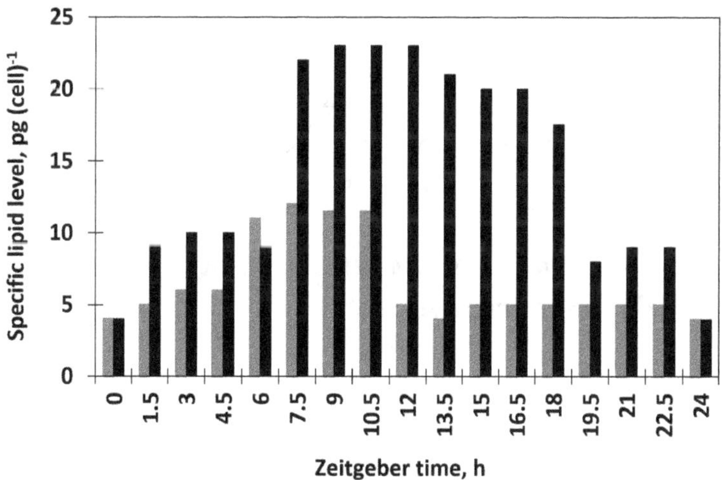

Figure 7.11. The specific lipid accumulation in a day beginning from the time the lights are switched on every day (Zeitgeber time). Gray bars represent control, and black bars represent the rhythm reset condition

7.8. Reactive Species in Improved Wastewater Treatment — Microwave and H_2O_2 Pretreatment (with Herald Wilson Ambrose)

Wastewater treatment is an important process, and methods to improve it contribute significantly to the overall human well-being. The treatment of wastewater at wastewater treatment plants (WWTPs) has limitations, such as excess residual sludge, pathogens, and odor at the early treatment stages. To overcome these limitations, anaerobic digestion is employed at WWTPs. During anaerobic digestion, the anaerobic microorganisms under anoxic conditions digest the activated sludge from the secondary treatment. The microbial consortium in the anaerobic digesters breaks down the complex organic matter through hydrolysis and subsequently metabolizes it through a series of biochemical reactions. Pretreatments that help solubilize the input-activated sludge can enhance such anaerobic digestions.

The pretreatments aim to weaken the complex network of extracellular polymeric substances and break down microbial cells to achieve solubilization. Some pretreatments result in the induction of intracellular RS, such as hydroxyl ($OH^{\bullet}$) and superoxide radicals ($O_2^{\bullet-}$), which help achieve microbial cell lysis. Besides, these free radicals also oxidize various organic pollutants in the sludge. Hence, the study on sludge solubilization with chemical oxidants gained increased attention in wastewater treatment. Furthermore, the impact of such oxidants on the activity of anaerobic microorganisms is crucial for subsequent anaerobic digestion.

7.8.1. *Induction of ROS in activated sludge by different pretreatments*

Hydrogen peroxide:

We used hydrogen peroxide, microwaves, and H_2O_2 + microwaves to effect the pretreatment before anaerobic digestion. Activated sludge, a source of various organic and inorganic constituents, mediates the reaction and leads to the solubilization of the complex matter. The generation of ($OH^{\bullet}$) and ($O_2^{\bullet-}$) radicals from H_2O_2 treatment through the Fenton reaction is given in the following equations [Ambrose *et al.*, 2019].

$$Fe^{2+} + H_2O_2 \rightarrow Fe^{3+} + OH^{\bullet} + HO^{-}$$

$$Fe^{3+} + H_2O_2 \rightarrow Fe^{2+} + HO_2^{\bullet} + H^{+}$$

$$OH^{\bullet} + H_2O_2 \rightarrow HO_2^{\bullet} + H_2O$$

$$HO_2^{\bullet} + OH^{\bullet} \rightarrow H_2O + O_2^{\bullet-}$$

The generation of specific intracellular ROS (siROS) in waste-activated sludge by hydrogen peroxide treatment was evaluated with fluorescent dyes such as aminophenyl fluorescein (APF) and DHE. The waste-activated sludge was collected from the aeration tank of Nesapakkam wastewater treatment plant in Chennai, India. The specific intracellular ROS levels (siROS) in the activated sludge for different treatments and the corresponding solubilization effect have been presented in Figure 7.12. Compared with 0.5% H_2O_2/TS (TS: total solids) treatment, 1% H_2O_2/TS treatment yielded higher levels of ($OH^{\bullet}$) radicals, as seen in Figure 7.12.

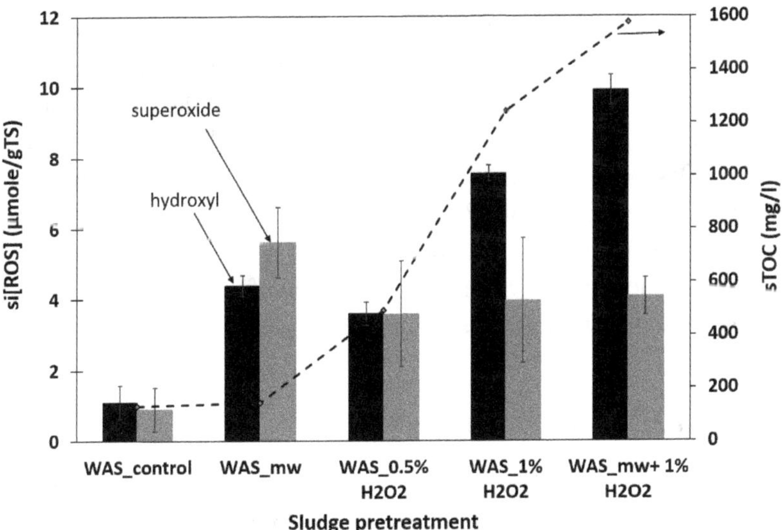

Figure 7.12. Effects of sludge pretreatments (microwave, H_2O_2, and combined) on intracellular ROS levels and waste-activated sludge solubilization. mw — microwave, WAS — waste activated sludge, H_2O_2 — hydrogen peroxide

However, the $(O_2^{\bullet-})$ levels were almost similar in both the treatments. Besides, the solubilization of sludge was studied by measuring soluble total organic carbon (sTOC). The variations in sTOC imply the degree of solubilized organic content achieved by the pretreatment in the sample. The increased flux of $(OH^{\bullet})$ radicals at a higher H_2O_2 concentration was concurrent with the increased sTOC concentration, as observed in Figure 7.12. Therefore, the disintegration of sludge and its microbial content can be attributed to the intracellular oxidative stress induced by H_2O_2 treatment.

Microwave

In the microwave (frequency range: 300 MHz–300 GHz; home microwave oven operates typically at 2.45 GHz) treatment, the electrically induced dipoles of the sample rapidly realign toward the oscillating electric field at high frequencies, resulting in the heating of the surrounding medium [Leonelli and Mason, 2010]. As the activated sludge is predominantly water, microwave irradiation can heat the sludge through the rapid realignment of water dipoles [Aylin Alagöz et al., 2016]. However, water evaporation is a challenge at high MW power, and the mechanism of sludge solubilization at lower microwave intensities has not been completely understood. Therefore, the microwave stress-induced generation of si[ROS] in the waste-activated sludge was studied. The microwave treatment at 450W power has shown the production of both $(OH^{\bullet})$ and $(O_2^{\bullet-})$ radicals, as shown in Figure 7.12. Thermal stress-induced production of siROS has been previously reported [Sarkar and Suraishkumar, 2011]. Producing intracellular ROS may be responsible for the sludge disintegration achieved at lower microwave intensities.

Combined H_2O_2 and microwave

The combination of different technologies was based on the synergism between individual methods. Different combinations have been studied to enhance sludge solubilization, dewaterability, and pathogen/odor reduction [Carrère et al., 2010; Gogate and Pandit, 2004]. The microwave and

H_2O_2 treatments exhibit synergism with each other as the dissociation of $(OH^{\bullet})$ radicals from H_2O_2 can be accelerated by microwave irradiation (Eswari *et al.*, 2016). Besides, the combined treatment technology can reduce the reaction time and chemical cost of the process. In the current study, the combination of 450 W microwave treatment with 1% H_2O_2 was studied for the solubilization of activated sludge. As seen in Figure 7.12, the generation of $(O_2^{\bullet-})$ radicals was significantly enhanced compared to the individual treatments. It was also accompanied by an increased solubilization of organics. The generation of siROS during mild treatment conditions may be crucial in choosing sludge pretreatments.

7.8.2. *Impact of pretreatment on oxidation status of anaerobic digestion*

After pretreatment, the activated sludge needs to be digested in anaerobic digesters for the production of biogas and the reduction of excess sludge in WWTPs. Anaerobic digestion involves a series of biochemical reactions, such as hydrolysis, acidogenesis, acetogenesis, and methanogenesis, carried out by anaerobic microbial communities [Meegoda *et al.*, 2018]. The interaction between these communities, consumption of substrates, and biogas formation occurs under anoxic conditions.

Anaerobic microorganisms have developed mechanisms to deal with ROS through enzymes such as superoxide reductases [Imlay, 2002]. In fact, the discovery of free radical neutralizing mechanisms in anaerobic microorganisms has extended the understanding of the evolution of "obligate anaerobiosis" on Earth [Ślesak *et al.*, 2019]. A vast majority of the obligate anaerobes carry genes for antioxidant response mechanisms [Imlay, 2008; Slesak *et al.*, 2012]. Hence, the evaluation of the "oxidation after-effects" of pretreatments on microbial activity becomes imminent to enhance the performance of anaerobic digestion.

We found relationships between specific superoxide levels and wastewater treatment parameters in this phase, such as sludge bioactivity and methane generation. The correlations were interesting [Ambrose *et al.*, 2019; Ambrose *et al.*, 2020] and provided insights into the workings of the anaerobic treatment process.

References

Ambrose, H. W., Philip, L., Sen, T. K., and Suraishkumar, G. K. (2019) The effect of combined microwave and hydrogen peroxide pretreatment on sludge characteristics and oxidation status of waste activated sludge. In *The Activated Sludge Process: Methods and Recent Developments*, (Lefèbvre, B. ed.), Nova Science Publishers, New York.

Ambrose, H. W., Philip, L., Suraishkumar, G. K., Karthikaichamy, A., and Sen, T. K. (2020). Anaerobic co-digestion of activated sludge and fruit and vegetable waste: evaluation of mixing ratio and impact of hybrid (microwave and hydrogen peroxide) sludge pre-treatment on two-stage digester stability and biogas yield. *J. Water Process Eng.*, 37, 101498.

Archanaa, S., Moise, S., and Suraishkumar, G. K. (2012). Chlorophyll interference in microalgal lipid quantification through the Bligh and Dyer method. *Biomass Bioenergy*, 46, 805.

Aylin Alagöz, B., Yenigün, O., and Erdinçler, A. (2016). Ultrasound assisted biogas production from co-digestion of wastewater sludges and agricultural wastes: comparison with microwave pre-treatment. *Ultrason. Sonochem.*, 40, 193.

Balan, R., and Suraishkumar, G. K. (2014 a). Simultaneous increases in specific growth rate and specific lipid content of *Chlorella vulgaris* through UV-induced reactive species. *Biotechnol. Prog.*, 30, 291.

Balan, R., and Suraishkumar, G. K. (2014 b). UVA-induced reset of hydroxyl radical ultradian rhythm improves temporal lipid production in *Chlorella vulgaris*. *Biotechnol. Prog.*, 30, 673.

Carrere, H., Antonopoulou, G., Affes, R., Passos, F., Battimelli, A., Lyberatos, G., and Ferrer, I. (2016). Review of feedstock pretreatment strategies for improved anaerobic digestion: from lab-scale research to full-scale application. *Bioresour. Technol.*, 199, 386.

Carrère, H., Dumas, C., Battimelli, A., Batstone, D. J., Delgenès, J. P., Steyer, J. P., and Ferrer, I. (2010). Pretreatment methods to improve sludge anaerobic degradability: a review. *J. Hazard. Mater.*, 183, 1.

Deitz, K. J., and Pfannschmidt, T. (2011). Novel regulators in photosynthetic redox control of plant metabolism and gene expression. *Plant Physiol.*, 155, 1477.

Deng, Y., and Zhao, R. (2015). Advanced oxidation processes (AOPs) in wastewater treatment. *Curr. Pollut. Rep.*, 1, 167.

Eskicioglu, C., Terzian, N., Kennedy, K. J., Droste, R. L., and Hamoda, M. (2007). Athermal microwave effects for enhancing digestibility of waste activated sludge. *Water Res.*, 41, 2457.

Eswari, P., Kavitha, S., Kaliappan, S., Yeom, I. T., and Banu, J. R. (2016). Enhancement of sludge anaerobic biodegradability by combined microwave-H2O2 pretreatment in acidic conditions. *Environ. Sci. Pollut. Res.*, 23, 13467.

Geigenberger, P., Kolbe, A., and Tiessen. (2005). A redox regulation of carbon storage and partitioning in response to light and sugars. *J Exp. Bot.*, 56, 1469.

Guarnieri, M. T., Nag, A., Smolinski, S. L., Darzins, A., Seibert, M., Pienkos, P. T. (2011). Examination of triacylglycerol biosynthetic pathways via de novo transcriptomic and proteomic analyses in an unsequenced microalga. *PLoS One*, 6, e25851.

Gogate, P. R., and Pandit, A. B. (2004). A review of imperative technologies for wastewater treatment II: hybrid methods, *Adv. Environ. Res.*, 8, 553.

Gonzalez, A., Hendriks, A. T. W. M., van Lier, J. B., and de Kreuk, M. (2018). Pre-treatments to enhance the biodegradability of waste activated sludge: elucidating the rate limiting step. *Biotechnol. Adv.*, 36, 1434.

Guan, R., Yuan, X., Wu, Z., Jiang, L., Li, Y., and Zeng, G. (2018). Principle and application of hydrogen peroxide based advanced oxidation processes in activated sludge treatment: a review. *Chem. Eng. J.*, 339, 519.

Imlay, J. A. (2008). How obligatory is anaerobiosis? *Mol. Microbiol.*, 68, 801.

Imlay, J. A. (2002). What biological purpose is served by superoxide reductase? *J. Biol. Inorg. Chem.*, 7, 659.

Leonelli, C., and Mason, T. J. (2010). Microwave and ultrasonic processing: now a realistic option for industry. *Chem. Eng. Process. Process Intensif.*, 49, 885.

Liochev, S. I. (2014). Free radicals: how do we stand them? Anaerobic and aerobic free radical (chain) reactions involved in the use of fluorogenic probes and in biological systems. *Med. Princ. Pract.*, 23, 195.

Liu, S., Wu, M., and Yao, X. (2020). Effects of reactive oxygen species scavengers on thermophilic micro-aerobic digestion for sludge stabilization. *Environ. Res.*, 185, 109453.

Meegoda, J. N., Li, B., Patel, K., and Wang, L. B. (2018). A review of the processes, parameters, and optimization of anaerobic digestion. *Int. J. Environ. Res. Public Health*, 15.

Menon, K. R., Balan, R., and Suraishkumar, G. K. (2013). Stress induced lipid production in *Chlorella vulgaris*: relationship with specific intracellular reactive species levels. 110, 1627.

Neyens, E., and Baeyens, J. (2003). A review of classic Fenton's peroxidation as an advanced oxidation technique. *J. Hazard. Mater.*, 98, 33.

Rismani-Yazdi, H., Haznedaroglu, B. Z., Hsin, C., and Peccia, J. (2012). Transcriptomic analysis of the oleaginous microalga *Neochloris oleoabundans* reveals metabolic insights into triacylglyceride accumulation. *Biotechnol. Biofuels*, 5, 74.

Sahoo, S., Verma, R., Suresh, A. K., Rao, K. K., Bellare, J., and Suraishkumar, G. K. (2003). Macro-level and genetic-level responses of *Bacillus subtilis* to shear stress. *Biotechnol. Prog.*, 19, 1689.

Sahoo, S., Rao, K. K., Suresh, A. K., and Suraishkumar, G. K. (2004). Intracellular reactive oxygen species mediate suppression of sporulation in *Bacillus subtilis* under shear stress. *Biotechnol. Bioeng.*, 87, 81.

Sahoo, S., Rao K. K., and Suraishkumar, G. K. (2006). Reactive oxygen species induced by shear stress mediate cell death in *Bacillus subtilis*. *Biotechnol. Bioeng.*, 94, 118.

Sarkar, P., and Suraishkumar, G. K. (2011). pH and temperature stresses in bioreactor cultures: intracellular superoxide levels. *Ind. Eng. Chem. Res.*, 50, 13129.

Ślesak, I., Kula, M., Ślesak, H., Miszalski, Z., and Strzałka, K. (2019). How to define obligatory anaerobiosis? An evolutionary view on the antioxidant response system and the early stages of the evolution of life on Earth. *Free Radic. Biol. Med.*, 140, 61.

Slesak, I., Slesak, H., and Kruk, J. (2012). Oxygen and hydrogen peroxide in the early evolution of life on earth: in silico comparative analysis of biochemical pathways. Astrobiol. 12, 775.

Suraishkumar, G. K. (2014). *Continuum Analysis of Biological Systems: Conserved Quantities, Forces and Fluxes*, Springer Publishing, Heidelberg.

Tyagi, V. K., and Lo, S. L. (2013). Microwave irradiation: a sustainable way for sludge treatment and resource recovery. *Renew. Sustain. Energy Rev.*, 18, 288.

Yang, Y., Zhang, C., and Hu, Z. (2013). Impact of metallic and metal oxide nanoparticles on wastewater treatment and anaerobic digestion. *Environ. Sci. Process. Impacts*, 15, 39.

Zhang, A., Wang, J., and Li, Y. (2015). Performance of calcium peroxide for removal of endocrine-disrupting compounds in waste activated sludge and promotion of sludge solubilization. *Water Res.*, 71, 125.

Zheng, L., Zhang, Z., Tian, L., Zhang, L., Cheng, S., Li, Z., and Cang, D. (2019). Mechanistic investigation of toxicological change in ZnO and TiO_2 multinanomaterial systems during anaerobic digestion and the microorganism response. *Biochem. Eng. J.*, 147, 62.

Chapter 8

Reactive Species Aspects in Nano-bio-systems

Nanotechnology is an important area that deals with materials fabricated at small length scales (particle sizes), called nanomaterials or nanoparticles. Typically, the particle size ranges from 1 to 100 nm [Caruso *et al.*, 2014]. The use of nanoparticles is sought-after for various industrial and commercial applications because the particles can be customized to have desired and enhanced or novel properties through size manipulation. The unique properties of nanoparticles are revolutionizing the capabilities of various products, including cosmetics, pharmaceuticals, textiles, paints, optics, sensors, catalysts, and so on [Biswas and Sarkar, 2019].

8.1. Nanotoxicity

Although nanoparticles are widely used, they can be toxic to cells of all life forms, including humans. The nanoparticles are internalized by cells through endocytosis. Once inside the cell, there are two major mechanisms by which nanoparticles cause toxicity:

1. oxidative stress because nanoparticles cause the induction of reactive species (RS), the molecular mediators of stress;
2. release of metal ions (say, Ag^+ from silver nanoparticles) due to nanoparticle dissolution.

With the information from the previous chapters, we can appreciate the reason for toxicity due to induced RS, the highly reactive molecules that can damage all fundamental biomolecules. Similarly, metal ions are toxic because they react with several important cellular molecules, especially proteins.

Some physicochemical factors determine the toxicity extent of the nanoparticles. Let us briefly describe some of them. For more specific details, please see Ahmed *et al.* [2017].

- *Size*: Nanoparticle size is a critical factor that influences toxicity. Typically, an inverse relationship exists between nanoparticle size and damage extent.
- *Shape*: The shape of the nanomaterial significantly determines the toxicity extent. For example, nanomaterials with cylindrical geometry (nano-wires) are more toxic than spherical nanoparticles.
- *Surface properties*: The surface properties that depend on the surface chemistry significantly determine the toxicity of the nanoparticles. The ease of ion release (say, Ag^+) from the surface directly determines toxicity. Surface charge and nature determine the extent of aggregation/agglomeration of nanoparticles, which, in turn, influence their toxicity.
- *Biocorona formation*: The nanoparticles interact with proteins, and proteins sorb onto nanoparticles to result in a complex called biocorona. The biocorona formation influences aspects such as uptake, stability, and ion dissolution. Thus, biocorona formation influences nanoparticle toxicity.

8.2. Nanotechnology is a Solution for Pollution but can be its Cause Too (with Archanaa Sundararaghavan)

Nanotechnology is possibly a solution for various problems that humanity faces [Aliev *et al.*, 2021]. The fact that nanotechnology has enormous potential to transform life is undeniable. However, the growing applications of nanoparticles and increasing production have raised concerns about their sustainable usage due to their possibility to harm the environment [Aliev *et al.*, 2021; Biswas and Sarkar, 2019]. From manufacture to disposal, the nanoparticles can enter the environmental media, such as water and soil, through various routes. The ecological exposure of nanoparticles is a potential risk to biological systems as their unique properties are not just a benefit but also tend to induce toxicity in the latter [Abbas *et al.*, 2020].

Due to their associated ecological toxicity risks, nanoparticles are emerging as a contaminant and have paved the way for novel pollution called nanopollution [Biswas and Sarkar, 2019]. Numerous studies have been carried out to understand the risks associated with the environmental release of nanoparticles and the mechanism behind their toxicity.

8.2.1. *RS: the molecular mediators of nanotoxicity*

As discussed in an earlier chapter, the surge in intracellular concentrations of RS, such as superoxide ($O_2^{\bullet-}$) and hydroxyl ($^{\bullet}OH$) radicals, was found to mediate the microalgal response to UV irradiation stress. It is also known that the RS surge resulting in oxidative stress influenced an organism's response to other abiotic stress factors such as pH, temperature, osmotic stress, altered nutrients, and so on [Wang *et al.*, 2004; Fan *et al.*, 2014] as we have seen in detail in the previous chapter. Not just the stress response, oxidative stress also influences the toxicity in biological systems towards various pollutants. For example, heavy metal induces toxic effects in biological systems by increasing the RS levels [Wu *et al.* 2016], as do pollutants like polyaromatic hydrocarbons [Kottuparambil and Park, 2019]. Thus, RS seem to be the ubiquitous mediators of stress in various cellular processes, and nanopollutants-induced toxicity is no exception.

The exposure to nanoparticles resulted in increased intracellular RS concentrations. As already mentioned in an earlier chapter, RS, when elevated above their physiological threshold concentrations, trigger various adverse effects in a biological system. By raising the RS levels, nanoparticles damage the macromolecules such as proteins, lipids, and DNA. The damage to these biomolecules disrupts the signal transduction pathway and the normal cellular function, eventually resulting in cytotoxicity, genotoxicity, and cell death [Fu *et al.*, 2014; Vakili-Ghartavol *et al.*, 2020].

The rapidly expanding research resulted in various engineered nanoparticles for commercial applications. Some examples are carbon-based nanoparticles, namely fullerene and carbon nanotubes (CNTs), silica, silver, and metal oxide nanoparticles like titanium dioxide (TiO_2), iron oxides (FeO_x), zinc oxide (ZnO), copper oxide (CuO) and cerium oxide (CeO) [Bundschuh *et al.*, 2018; Biswas and Sarkar, 2019]. All these nanoparticles can induce toxicity via oxidative stress resulting from increased intracellular RS concentrations (Table 8.1).

Table 8.1. Toxic effects of different nanoparticles on various model systems

Nanoparticles	Ecosystem/model organism	Observations
Fullerene	Aquatic/freshwater fish (*Pseudetroplus maculatus*) [Sumi and Chitra, 2017]	Oxidative stress (Increased H_2O_2 levels) — altered metabolism in muscle tissue
	Aquatic/*Daphia magna* [Lv *et al.*, 2017]	Oxidative stress (Increased lipid peroxidation) — Gut impairment, inhibition of digestive enzymes, altered mobility
Carbon nanotubes	Terrestial/*Arabidopsis thaliana* [Shen *et al.*, 2010]	Oxidative stress (H_2O_2 accumulation) — adverse effects on protoplast and leaves, programmed cell death
	Aquatic/*Chlorella pyrenoidosa* [Zhang *et al.*, 2015]	Oxidative stress (Increased total RS and malondialdehyde [MDA] levels) — Growth inhibition
	Aquatic/freshwater fish(*Channa punctatus*) [Ali *et al.*, 2020]	Oxidative stress (increased lipid peroxidation) — Genotoxicity
Silver	Terrestrial/wheat [Dimkpa *et al.*, 2013]	Oxidative stress (oxidized glutathione accumulation) — adverse effects on plant metabolism and growth
	Terrestial/Earthworm (*Eisenia fetida*) [Das *et al.*, 2018]	Oxidative stress (increased lipid peroxidation) — reduced protein synthesis, significant weight loss

Table 8.1.   (*Continued*)

Nanoparticles	Ecosystem/model organism	Observations
Titanium dioxide	*Caenorhabditis elegans* [Sonane, *et al.*, 2017]	Oxidative stress (increased total RS levels) — increased mortality
	Terrestrial/land snail (*Cornu aspersum*) Aquatic/zebrafish (*Danio rerio*), and the Prussian carp (*Carassius gibelio*) [Bobori *et al.*, 2020]	Oxidative stress (increased superoxide levels and lipid peroxidation) — lysosomal membrane disruption, protein oxidation, and genotoxicity were observed in all three animal models. Zebrafish displayed decreased swimming ability
Iron oxide	Aquatic/mussel (*Mytilus galloprovincialis*) [Taze *et al.*, 2016]	Oxidative stress (increased total RS levels) — DNA damage, protein oxidation
	Terrestrial/land snail (*Cornu aspersum*) Aquatic/zebrafish (*Danio rerio*), and the prussian carp (*Carassius gibelio*) [Kaloyianni *et al.*, 2020]	Oxidative stress (increased lipid peroxidation) — genotoxicity were observed in all three animal models. Zebrafish displayed decreased swimming ability
Zinc oxide	Aquatic/zebrafish (*Danio rerio*) [Zhao *et al.*, 2016]	Oxidative stress (increased lipid peroxidation) — DNA damage, apoptosis, increased heart rate, decreased rate of embryo hatching
	Caenorhabditis elegans [Sonane, *et al.*, 2017]	Oxidative stress (increased total RS levels) — increased mortality
	Terrestrial/Silkworm (*Bombyx mori*) [Mir *et al.*, 2020]	Oxidative stress (increased superoxide and H_2O_2 levels) — decreased hemocyte viability, altered morphology, apoptotic cell death
Copper oxide	Aquatic/juvenile carps (*Cyprinus carpio*) [Gupta *et al.*, 2016]	Oxidative stress (increased antioxidant enzyme activities) — disrupted secondary lamella of gill, liver damage, degenerated renal tubules in the kidney
	Terrestrial/Tomato (*Solanum lycopersicon*) [Ahmed, *et al.*, 2018]	Oxidative stress (increased total RS and MDA levels) — decreased growth, dry biomass and total protein, cellular damage

Table 8.1.   (*Continued*)

Nanoparticles	Ecosystem/model organism	Observations
	Terrestrial/Field mustard (*Brassica rapa*) [Chung *et al.*, 2019]	Oxidative stress (increased total RS, H_2O_2 and MDA levels) — DNA damage, decreased chlorophyll and carotenoid content, reduced shoot, and root length
Cerium oxide	Terrestrial/Tomato (*Solanum lycopersicum* L.) [Wang *et al.*, 2013]	Oxidative stress (increased H_2O_2 levels) — reduction in biomass, decreased water transpiration
	Aquatic/amphipods (*Corophium volutator*) [Dogra *et al.*, 2016]	Oxidative stress (Increased lipid peroxidation) — DNA damage [single strand DNA breaks]
	Aquatic/crustacean (*Artemia salina*) [David *et al.*, 2017]	Oxidative stress (Increased total RS) — inhibited hatching, increased mortality

8.2.2. *Nanoparticle-induced RS generation*

Nanoparticles can generate RS in biological systems by various means. As mentioned in an earlier section, nanoparticles are prone to internalization by the cell. Inside the cell, some metal and metal–oxide nanoparticles can undergo dissolution, resulting in the release of their respective metal ion [Bundschuh *et al.*, 2018; Dayem *et al.*, 2017]. The released metal ion can then undergo a Fenton-like reaction to generate the most reactive form of RS, namely the hydroxyl ($^\bullet OH$) radical. For example, the copper ions from CuO nanoparticle react with H_2O_2 (Fenton-like reaction) to generate $^\bullet OH$ [Yin *et al.*, 2012], while the FeO_x nanoparticles generate RS through Fenton reaction [Dayem *et al.*, 2017].

Let us recall that the Fenton–like reaction of copper ions is

$$Cu^+ + H_2O_2 \rightarrow Cu^{2+} + OH^- + {}^\bullet OH$$

The copper ions can also generate other RS, namely superoxide radical and hydrogen peroxide, by reacting with molecular oxygen as follows [Wang *et al.*, 2010].

$$Cu^+ + O_2 \rightarrow Cu^{2+} + O_2^{\cdot-}$$

$$Cu^{2+} + O_2^{\cdot-} + 2H^+ \rightarrow Cu^{2+} + H_2O_2$$

The generated superoxide radical and H_2O_2 can produce hydroxyl radicals through the Haber–Weiss reaction [Fu *et al.*, 2014], a superoxide-powered Fenton reaction.

$$O_2 \qquad\qquad O_2^{\cdot-}$$

$$H_2O_2 + Fe^{2+} \longrightarrow Fe^{3+} + {}^{\bullet}OH + OH^- \text{ (Fenton reaction)}$$

Like copper ions from CuO nanoparticles, the silver nanoparticles also generate hydroxyl radicals through the Fenton-like reaction, as given below. Silver ion also reacts with molecular oxygen to generate superoxide radicals [Fu *et al.*, 2014].

$$Ag + H_2O_2 \rightarrow Ag^+ + {}^{\bullet}OH + OH^-$$

Thus, with nanoparticles that can dissolve to release their respective metal ions, the generation of RS is predominantly via Fenton and Haber–Weiss reaction. However, some photocatalytic nanoparticles like TiO_2 are difficult to dissolve. Yet, TiO_2 can generate RS via light (UV or visible) induction due to its superconductive properties.

When TiO_2 is excited with light photons, a positive hole (h^+) in the valence band and a conduction band electron (e^-) result.

$$TiO_2 \xrightarrow{\ h\upsilon\ } h^+ + e^-$$

The positive hole and the electron react with molecular oxygen and water to generate RS [Mathur *et al.*, 2015].

$$h^+ + H_2O \rightarrow {}^{\bullet}OH + H^+$$

$$e^- + O_2 \rightarrow O_2^{\bullet-}$$

$$2O_2^{\bullet-} + 2\,H_2O \rightarrow H_2O_2 + 2OH^- + O_2$$

$$e^- + H_2O_2 \rightarrow OH^- + {}^\bullet OH$$
$$h^+ + OH^- \rightarrow {}^\bullet OH$$

Another photocatalytic nanoparticle, ZnO, also generates RS via a process similar to that of TiO_2 [Fu *et al.*, 2014]. While the generation of RS by TiO_2 nanoparticles is believed to be light-assisted, they can induce oxidative stress even in the absence of light. For example, when the bacteria *Rhodococcus opacus* PD630 was exposed to TiO_2 nanoparticles under dark conditions, a significant increase in specific intracellular hydroxyl radical (siOH) levels was observed [Sundararaghavan *et al.*, 2019]. The percentage increase in siOH levels w.r.t control proved the induction of oxidative stress, and the stress intensity increased with increasing nanoparticle concentration (Figure 8.1).

The induction of oxidative stress by TiO_2 under dark conditions suggests that their environmental exposure could mean harm to darker zone biotics as well [Zhang *et al.*, 2017]. While the exact mechanism behind the rise in RS levels under dark conditions is unknown, the nanoparticle might work by activating the RS-associated enzymes such as NADPH oxidase [Dayem *et al.*, 2017].

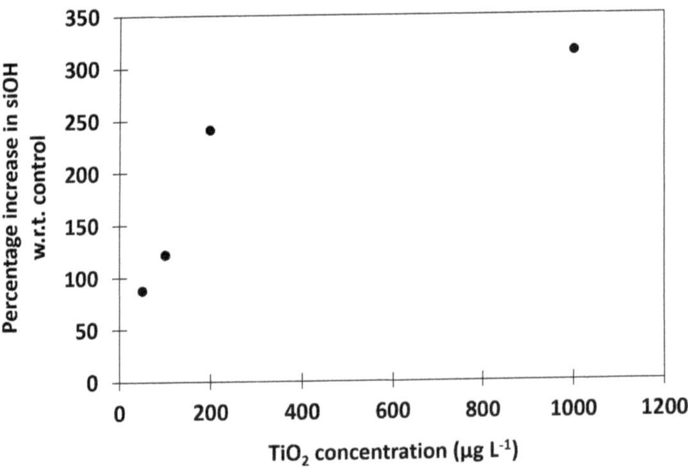

Figure 8.1. Induction of oxidative stress by TiO_2 nanoparticles under dark conditions. The oxidative stress is inferred from the percentage increase in specific intracellular hydroxyl radical (siOH) level with respect to control

8.2.3. *A positive aspect of nanoparticle-induced oxidative stress: a role in managing nanopollution*

With nanoparticles' expanding applications, their global production capacity is increasing rapidly, and so are the concerns related to their release and undesirable effects on the environment [Biswas and Sarkar, 2019; Nasrollahzadeh *et al.*, 2019]. Whether nanotechnology is a solution or a threat to humanity is highly debatable. Though nanopollution is an emerging concern, the benefits associated with nanoparticles are enormous, and so it is hard to hamper their application or production. However, controlling their accidental release into the environment seems to be a viable approach.

One such approach has been proposed for controlling the environmental release of TiO_2, the most widely produced nanoparticle for various commercial purposes. The approach involved using oleaginous bacteria *R. opacus* PD630 to recover the TiO_2 nanoparticles from the municipal wastewater treatment plant effluent, which may otherwise form a significant point for nanoparticle entry into the environment [Sundararaghavan *et al.*, 2019]. The approach seemed promising, as the proof of concept showed that the bacteria could recover TiO_2 nanoparticles from the water spiked with 1000 PPB of TiO_2. The nanoparticle recovery efficiency of the bacteria improved further with UV light assistance. With UV irradiation throughout the growth phase, the bacteria's nanoparticle recovery efficiency improved. When UV irradiation was initiated from the mid-log phase of the growth, the recovery efficiency improved significantly to 97%. Overall, the study demonstrated the bacteria *R. opacus* PD 630's potential to recover the TiO_2 nanoparticles from the waste stream before its release, thereby controlling the environmental release of nanoparticles and thus ensuring their sustainable use.

The high TiO_2 recovery potential of *R. opacus* PD630 could be due to the presence of a protein with a high affinity for TiO_2, namely the titanium-binding protein (TiBP). The TiBP, which has an exceptionally high affinity for TiO_2 nanoparticles, is a cell surface homolog of cytosolic dihydrolipoyl dehydrogenase (DLDH) and was first discovered to be present in *R. ruber* [Gertler *et al.*, 2003]. One of the protein sequences of *R. opacus* PD630 (annotated as DLDH) had a sequence identity of 90% with the

```
R.opacus_prot     51 KALLRNAELAHLFTKEAKTFGISGEASFDFGAAYDRSRKVADGRVKGVHF    100
                     ||||||||||||||||||||||:||:.|||||||.|||||||||||||||
R.ruber_TiBP      51 KALLRNAELAHLFTKEAKTFGMSGDVSFDFGAALDRSRKVADGRVKGVHF    100

R.opacus_prot    101 LMKKNKITEYDGKGSFTDANTLSVELSKGGTETVTFDNAIIATGSTTKLL    150
                     |||||||||||||.|||.||||:||||:|||||||||||||||||.||||
R.ruber_TiBP     101 LMKKNKITEYDGAGSFVDANTISVELNKGGTETVTFDNAIIATGSVTKLL    150

R.opacus_prot    151 PGTSLSENVVTYEEQILTRDLPESILIVGAGAIGMEFGYVLKNYGVDVTI    200
                     |||.|||||||||||||:||.||||||||||||||||||||||||||||||
R.ruber_TiBP     151 PGTQLSENVVTYEEQILTRELPGSILIVGAGAIGMEFGYVLKNYGVDVTI    200

R.opacus_prot    201 VEFLDRALPNEDADVSKEIEKQYKKLGVTIKTGAAVQSIDDDGSKVTVSI    250
                     ||||||||||||||:||||.|.||||||.|.||||||||||||:||||:|
R.ruber_TiBP     201 VEFLDRALPNEDADISKEIAKAYKKLGVKIITGAAVQSIDDDGNKVTVAI    250

R.opacus_prot    251 KNNKSGDIETVVVDKVMQSVGFAPRVEGFGLEKTGVQL-DRGAIGITDTM    299
                     |:||||::|:||||||||||||||||||||||||.|||:| |||||.|.|.|
R.ruber_TiBP     251 KDNKSGNVESVVVDKVMQSVGFAPRVEGFGLENTGVELTDRGAIAIDDVM    300

R.opacus_prot    300 QTSVPHIYAIGDVTMKLQLAHVAEAQGVVAAETIAGVETLPIEDYRMMPR    349
                     :|:|||||||||||.||.|||||||.|||||||.|.|||..:|||||||
R.ruber_TiBP     301 RTNVPHIYAIGDVTAKLMLAHVAEAMGVVAAETIGGAETLTFDDYRMMPR    350

R.opacus_prot    350 ATFCQPQVASFGLTEQQAKDEGYDVKVATFPFTANGKAHGLGDPTGFVKL    399
                     |||||||||||||||||:|||||||||||||||||||||||||.|||||
R.ruber_TiBP     351 ATFCQPQVASFGLTEQQARDEGYDVKVATFPFTANGKAHGLGDPNGFVKL    400

R.opacus_prot    400 IADKKYGELLGGHLIGPDVSELLPELTLAQKWDLTVNELARNVHTHPTLS    449
                     |||.|:||||||||||||||||||||||||||||||||||||||||||||
R.ruber_TiBP     401 IADNKHGELLGGHLIGPDVSELLPELTLAQKWDLTVNELARNVHTHPTLS    450

R.opacus_prot    450 EALQEAIHGLAGHMINF       466
                     |||||||||||||||||
R.ruber_TiBP     451 EALQEAIHGLAGHMINF       467
```

Figure 8.2. Pairwise alignment of *R. ruber*'s TiBP and the DLDH protein of *R. opacus* PD630

TiBP sequence of *R. ruber,* and the two sequences aligned perfectly (Figure 8.2). The perfect sequence alignment indicates the functional homology between the two sequences [Koonin and Galperin, 2003]. Further, the docking score and the atomic contact energy (ACE) obtained using patchdock of *R. opacus*' protein, and TiO_2 nanoparticle complex was comparable to that of *R. ruber*'s TiBP and TiO_2 nanoparticle complex (Figure 8.3). The identical score and similar ACE (binding energy) show that, like *R. ruber, R. opacus* PD630 also exhibits a high affinity for TiO_2 nanoparticles, which explains their high TiO_2 recovery potential, discussed previously.

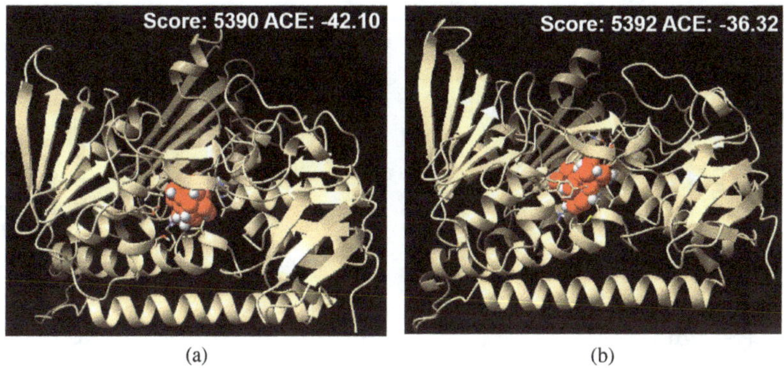

Figure 8.3.   The docking score and ACE of *R. ruber*'s TiBP (a) and *R. opacus* PD630's DLDH (b) when docked with TiO$_2$ nanoparticle

In addition to TiO$_2$ nanoparticles recovering potential, the bacteria *R. opacus* PD630 also accumulates a significant amount of triacylglycerol (TAG) — the precursor of biodiesel [Sundararaghavan *et al.*, 2019] and here comes the role of nanoparticle-induced oxidative stress. As mentioned before, exposure of *R. opacus* PD630 to TiO$_2$ nanoparticles induced oxidative stress in the bacteria (Figure 8.1). Oxidative stress has been known to influence the TAG content of the oleaginous organisms positively, and few such instances have been discussed in the previous chapters. Likewise, the TiO$_2$ nanoparticle-induced oxidative stress enhanced the TAG accumulation in *R. opacus* PD630 [Sundararaghavan *et al.*, 2019]. As shown in Figure 8.4, as the oxidative stress in the bacteria increased with increasing TiO$_2$ nanoparticle concentration, the TAG content of the bacteria also increased simultaneously. While the above trend was observed under the presence of UV light irradiation initiated from the mid-log phase, the trend was similar in the other cases, too, that is, the absence of and continuous UV light irradiation [Sundararaghavan *et al.*, 2019].

Thus, the approach proposed to control the environmental release of TiO$_2$ nanoparticles [Sundararaghavan, *et al.*, 2019] brings in the advantage of oxidative stress to double up the purpose, that is, recovery of TiO$_2$ nanoparticles from the waste stream using the bacteria and concurrent increase in production of industrially valuable TAG by the bacteria. The strategy is reasonable for other nanoparticles as well.

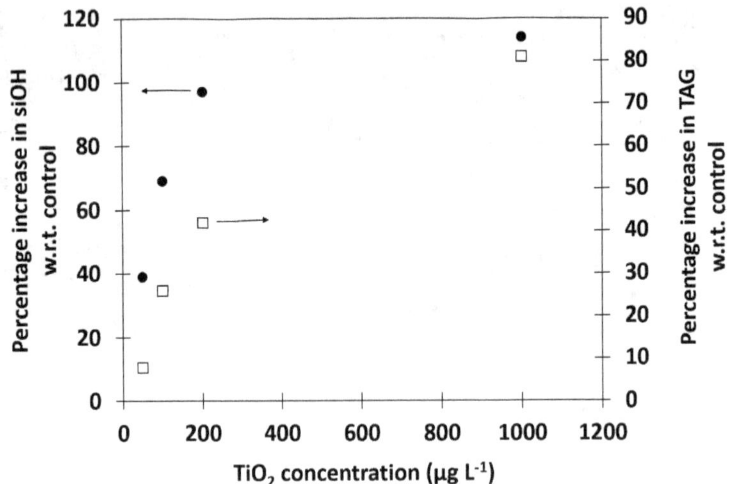

Figure 8.4. Percentage increase in siOH levels and TAG content w.r.t control in *R. opacus* PD630 exposed to different TiO_2 nanoparticle concentrations

As already mentioned, the enormous benefits associated with nano-technology overshadow their anticipated ecological risks. However, if the release of nanoparticles into the environment is ignored now, they might evolve as a significant threat to the environment due to their oxidative stress-inducing nature. When the development of a regulatory framework for nanoparticle usage is progressing gradually, research, in parallel, on controlling their environmental levels is essential.

References

Abbas, Q., Yousaf, B., Amina, Ali, M. U., Munir, M. A. M., El-Naggar, A., Rinklebe, J., and Naushad. M. (2020). Transformation pathways and fate of engineered nanoparticles (ENPs) in distinct interactive environmental compartments: a review. *Environ. Int..* 138, 105646.

Ahmed, K. B. R., Nagy, A. M., Brown, R. P., Zhang, Q., Malghan, S. G., and Goering, P. L. (2017). Silver nanoparticles: significance of physicochemical properties and assay interference on the interpretation of in vitro cytotoxicity studies. *Toxicol. in vitro*, 38, 179.

Ahmed, B., Khan, M. S., and Musarrat, J. (2018). Toxicity assessment of metal oxide nano-pollutants on tomato (*Solanum lycopersicon*): a study on growth dynamics and plant cell death. *Environ, Poll.*, 240, 802.

Ali, D., Falodah, F. A., Almutairi, B., Alkahtani, S., and Alarifi, S. (2020). Assessment of DNA damage and oxidative stress in juvenile *Channa punctatus* (bloch) after exposure to multi-walled carbon nanotubes. *Environ. Toxicol.*, 35, 359.

Aliev, B. H., Kazymov, N. F., Aliev, Z. H., Akhmedova, G. M., Feyzullaeva, K. S., Ismailova, L. M., and Galandarov, G. A. (2021). Nanotechnology as one of the possible solutions to the problem of civilization or threat to humanity. *Int. J Pharm. Inft. Thrp.*, 4, 1.

Biswas, J. K., and Sarkar, D. (2019). Nanopollution in the aquatic environment and ecotoxicity: no nano issue! *Curr. Pollution Rep.*, 5, 4.

Bobori, D., Dimitriadi, A., Karasiali, S., Tsoumaki-Tsouroufli, P., Mastora, M., Kastrinaki, G., Feidantsis, K., Printzi, A., Koumoundouros, G., and Kaloyianni, M. (2020). Common mechanisms activated in the tissues of aquatic and terrestrial animal models after TiO$_2$ nanoparticles exposure. *Environ. Int.*, 138, 105611.

Bundschuh, M., Filser, J., Lüderwald, S., McKee, M. S., Metreveli, G., Schaumann, G. E., Schulz, R. and Wagner, S. (2018). Nanoparticles in the environment: where do we come from, where do we go to? *Environ. Sci. Europe*, 30, 1.

Caruso, G., Merlo, L., and Caffo, M. (2014). *Innovative Brain Tumor Therapy: Nanoparticle Based Therapeutic Strategies*. Woodhead Publishing, Sawston.

Chung, I. M., Rekha, K., Venkidasamy, B., and Thiruvengadam, M. (2019). Effect of copper oxide nanoparticles on the physiology, bioactive molecules, and transcriptional changes in Brassica rapa ssp. rapa seedlings. *Water, Air, Soil Poll.*, 230, 48.

Das, P., Barua, S., Sarkar, S., Chatterjee, S. K., Mukherjee, S., Goswami, L., Das, S., Bhattacharya, S., Karak, N., and Bhattacharya, S. S. (2018). Mechanism of toxicity and transformation of silver nanoparticles: inclusive assessment in earthworm-microbe-soil-plant system. *Geoderma*, 314, 73.

Dayem, A. A., Hossain, M. K., Lee, S. B., Kim, K., Saha, S. K., Yang, G., Choi, H. Y., and Cho, S. (2017). The role of reactive oxygen species (ROS) in the biological activities of metallic nanoparticles. *Int. J. Mol. Sci.*, 18, 120.

Dimkpa, C. O., McLean, J. E., Martineau, N., Britt, D. W., Haverkamp, R., and Anderson, A. J. (2013). Silver nanoparticles disrupt wheat (*Triticum aestivum* L.) growth in a sand matrix. *Environ. Sci. Tech.*, 47, 1082.

Dogra, Y., Arkill, K. P., Elgy, C., Stolpe, B., Lead, J., Valsami-Jones, E., Tyler, C. R., and Galloway, T. S. (2016). Cerium oxide nanoparticles induce oxidative stress in the sediment-dwelling amphipod *Corophium volutator*. *Nanotoxicol.*, 10, 480.

Fan, J., Cui, Y., Wan, M., Wang, W. and Li, Y. (2014). Lipid accumulation and biosynthesis genes response of the oleaginous Chlorella pyrenoidosa under three nutrition stressors. *Biotechnol. Biofuels,* 7, 17.

Fu, P. P., Xia, Q., Hwang, H. M., Ray, P. C., and Yu, H. (2014). Mechanisms of nanotoxicity: generation of reactive oxygen species. *J Food Drug Anal.,* 22, 64.

Gertler, G., Brudo, I., Kenig, R., and Fleminger, G.. (2003). A TiO$_2$-binding protein isolated from Rhodococcus strain GIN-1 (NCIMB 40340) — purification, properties and potential applications. *Materialwissenschaft Werkstofftechnik,* 34, 1138.

Gupta, Y. R., Sellegounder, D., Kannan, M., Deepa, S., Senthilkumaran, B., and Basavaraju, Y. (2016). Effect of copper nanoparticles exposure in the physiology of the common carp (Cyprinus carpio): biochemical, histological and proteomic approaches. *Aquacult. Fish.,* 1, 15.

Kaloyianni, M., Dimitriadi, A., Ovezik, M., Stamkopoulou, D., Feidantsis, K., Kastrinaki, G., Gallios, G. Tsiaoussis, I., Koumoundouros, G., and Bobori, D. (2020). Magnetite nanoparticles effects on adverse responses of aquatic and terrestrial animal models. *J. Hazard. Mater.* 383, 121204.

Koonin, E. V, and Galperin, M. Y. (2003) Evolutionary concept in genetics and genomics. In *Sequence–Evolution–Function: Computational Approaches in Comparative Genomics.* Kluwer Academic Publications, Boston.

Kottuparambil, S., and Park. J. (2019). Anthracene phytotoxicity in the freshwater flagellate alga Euglena agilis carter. *Sci. Rep.,* 9, 15323.

Lv, X., Huang, B., Zhu, X., Jiang, Y., Chen, B., Tao, Y., Zhou, J., and Cai, Z. (2017). Mechanisms underlying the acute toxicity of fullerene to Daphnia magna: energy acquisition restriction and oxidative stress. *Water Res.,* 123, 696.

Mathur, A., Raghavan, A., Chaudhury, P., Johnson, J. B., Roy, R., Kumari, J., Chaudhuri, G., Chandrasekaran, N., Suraishkumar, G. K., and Mukherjee, A. (2015). Cytotoxicity of titania nanoparticles towards waste water isolate *Exiguobacterium acetylicum* under UVA, visible light and dark conditions. *J. Environ. Chem. Eng.,* 3, 1837.

Mir, A. H., Qamar, A., Qadir, I., Naqvi, A. H., and Begum, R. (2020). Accumulation and trafficking of zinc oxide nanoparticles in an invertebrate model, *Bombyx mori,* with insights on their effects on immuno-competent cells. *Sci. Rep.,* 10, 1617.

Nasrollahzadeh, M., Sajadi, M. S., Sajjadi, M., Issaabadi, Z., and Atarod, M. (2019) *An Introduction to Green Nanotechnology,* Academic Press, Amsterdam.

Shen, C. X., Zhang, Q. F., Li, J., Bi, F. C., and Yao, N. (2010). Induction of programmed cell death in Arabidopsis and rice by single-wall carbon nanotubes. *Am. J. Bot.*, 97, 1602.

Sonane, M., Moin, N., and Satish, A. (2017). The role of antioxidants in attenuation of *Caenorhabditis elegans* lethality on exposure to TiO_2 and ZnO nanoparticles. *Chemosphere*, 187, 240.

David, E. M. D. S., Royam, M. M., Sekar, S. K. R., Manivannan, B., Soman, S. J., Mukherjee, A., and Natarajan, C. (2017). Toxicity, uptake, and accumulation of nano and bulk cerium oxide particles in *Artemia salina*. *Environ. Sci. Poll. Res.*, 24, 24187.

Sumi, N., and Chitra, K. C. (2017). Oxidative stress in muscle tissue of the freshwater fish, *Pseudetroplus maculatus* (bloch 1795): a toxic response from exposure to fullerene (C60) nanoparticles. *Asian Fish. Sci.*, 30, 206.

Sundararaghavan, A., Mukherjee, A., and Suraishkumar, G. K.. (2019). Investigating the potential use of an oleaginous bacterium, *Rhodococcus opacus* PD630, for nano-TiO_2 remediation. *Environ. Sci. Poll. Res.*, 27, 27394.

Taze, C., Panetas, I., Kalogiannis, S., Feidantsis, K., Gallios, G. P., Kastrinaki, G. P., Konstandopoulos, A. G., Václavíková, M., Ivanicova, L., and Kaloyianni, M. (2016). Toxicity assessment and comparison between two types of iron oxide nanoparticles in Mytilus galloprovincialis. *Aquat. Toxicol.*, 172, 9.

Vakili-Ghartavol, R., Momtazi-Borojeni, A. A., Vakili-Ghartavol, Z., Aiyelabegan, H. T., Jaafari, M. R., Rezayat, S. M., and Bidgoli, S. A. (2020). Toxicity assessment of superparamagnetic iron oxide nanoparticles in different tissues. *Artif. Cells, Nanomed. Biotechnol.*, 48, 443.

Wang, F., Jiao, P., Qi, M., Frezza, M., Dou, Q. P., and Yan, B. (2010). Turning tumor-promoting copper into an anti-cancer weapon via high-throughput chemistry. *Curr. Med. Chem.*, 17, 2685.

Wang, Q., Ebbs, S. D., Chen, Y., and Ma, X. (2013). Trans-generational impact of cerium oxide nanoparticles on tomato plants. *Metallomics*, 5, 753.

Wang, S. B., Chen, F., Sommerfeld, M., and Hu, Q. (2004). Proteomic analysis of molecular response to oxidative stress by the green alga *Haematococcus pluvialis* (Chlorophyceae). *Planta*, 220, 17.

Wu, X., Cobbina, S. J., Mao, G., Xu, H., Zhang, Z., and Yang, L. (2016). A review of toxicity and mechanisms of individual and mixtures of heavy metals in the environment. *Environ. Sci. Poll. Res.*, 23, 8244.

Yin, J. J., Fu, P. P., Lutterodt, H., Zhou, Y. T., Antholine, W. E., and Wamer, W. (2012). Dual role of selected antioxidants found in dietary supplements:

crossover between anti- and pro-oxidant activities in the presence of copper. *J. Agric. Food Chem.*, 60, 2554.

Zhang, L., Lei, C., Chen, J., Yang, K., Zhu, L., and Lin, D. (2015). Effect of natural and synthetic surface coatings on the toxicity of multiwalled carbon nanotubes toward green algae. *Carbon*, 83, 198.

Zhang, Z., Zhang, J., Shi, C., Guo, H., Ni, R. Y., Qu, J., Tang, J., and Liu, S. (2017). Effect of oxidative stress from nanoscale TiO_2 particles on a *Physarum polycephalum* macroplasmodium under dark conditions. *Environ. Sci. Poll. Res.*, 24, 17241.

Zhao, X., Ren, X., Zhu, R., Luo, Z., and Ren, B. (2016). Zinc oxide nanoparticles induce oxidative DNA damage and ROS-triggered mitochondria-mediated apoptosis in zebrafish embryos. *Aquat. Toxicol.,* 180, 56.

Chapter 9

Reactive Species in Cancer Treatment and Management

In Chapter 1, it was hinted that reactive species (RS) play an important role in cancer. In fact, RS play two diametrically opposite roles in cancer. At high enough levels above the homeostatic levels, they cause cancer — they turn normal cells into cancerous ones. But, at much higher levels, they are used to kill cancerous cells, and thus they treat/ manage cancer.

As mentioned in earlier chapters, many environmental factors such as radiation, heavy metals, drugs, cigarette smoke, some chemicals, and many pollutants can cause RS induction in cells. The induced RS can affect the fundamental biomolecules such as proteins, lipids, and nucleic acids. The damage can lead to the induction of various cellular signaling cascades that cause the activation of oncogenes and inactivation of tumor-suppressor genes, which, in turn, lead to tumor formation (tumorigenesis). Such environmental factors are, thus, carcinogenic (cancer-causing).

The RS level in cancer cells is higher than in normal cells. However, at high enough RS levels, any cell, whether cancerous or normal, can be killed. Thus, if high enough RS levels are induced in cancer cells, they can be killed. Indeed, this is the fundamental basis for the action of chemotherapy and radiotherapy in cancer treatment. The treatment kills cancerous cells and does not kill normal cells (Figure 9.1).

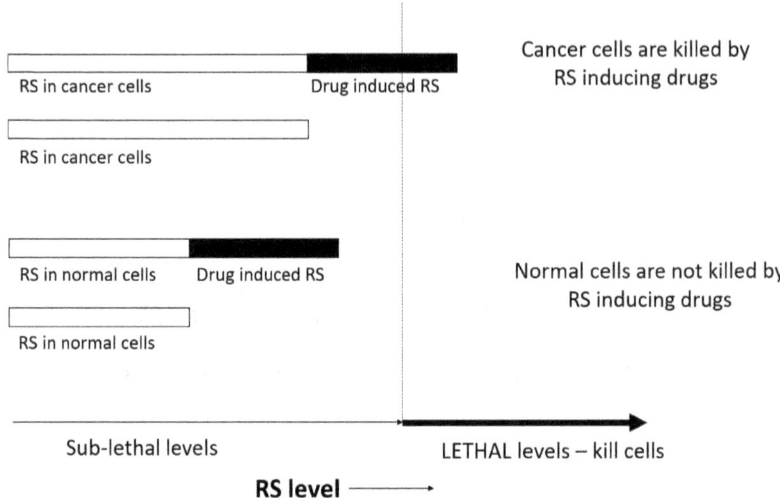

Figure 9.1.   The total RS level in the presence of RS-inducing drugs is higher in cancer cells than normal cells because the basal RS levels are higher in cancer cells. At appropriately high total RS levels, the cancer cells can be killed without the normal cells being killed

9.1. RS can Promote Cancer Cell Formation as well as Kill Cancer Cells

There is significant evidence in the literature (see Galadari *et al.*, [2017] for a review) that RS can have opposite effects — depending on the level, they can either promote cancer (pro-tumor effect) or kill cancer cells (anti-tumor effect). RS promote the following pro-tumor events:

–   Tumorigenesis: The RS-caused DNA damage can activate oncogenes (cancer-causing genes) or inactivate tumor suppressor genes. Both these actions promote cancer formation. RS can also activate various intracellular signaling cascades that lead to cancer formation. In addition, RS cause tumorigenesis by other molecular means, too.

–   Angiogenesis: New blood vessels are formed from existing capillaries to support rapid tumor growth. This process is called angiogenesis. RS regulate the processes that result in angiogenesis, namely, endothelial cell formation, migration, and tube formation.

— Metastasis: Cancer is deadly in 90% of the cases because it can move from the primary location through the bloodstream and lodge itself in a new location in the body. This movement of cancer cells is called metastasis. The entry of cancer cells into the bloodstream is called intravasation, and their exit from the bloodstream is called extravasation. RS are involved in various molecular events that determine intravasation, metastasis, and extravasation.

In addition, RS are involved in chemoresistance [Kim *et al.*, 2019] or the ineffectiveness of existing drugs to treat cancer.

RS also promote the following antitumor events, which are different active cell death mechanisms, depending on their intracellular level: Apoptosis, autophagy, necroptosis, and ferroptosis.

In the remaining sections, let us look at some interesting aspects of RS in the context of cancer treatment and management.

9.2. RS and Redox Ratio in a Microbiome Organism in a Cancer Context

It is now known that the gut microbiome plays a significant role in cancer. The gut microbiota synthesize molecules such as folate, which play a crucial role in preventing cancer onset, especially colorectal cancer. In addition, the gut microbiota are involved in regulating the host's antioxidant responses. Further, the gut microbiome also determines the effectiveness of traditional cancer therapy in humans and are suspected to play a role in cancer immunotherapy effectiveness by modulating the host's response to the recent therapy. We found some interesting phenomena related to gut microbiota and RS levels, which are described in this section.

As mentioned in Chapter 1, traditional cancer therapies such as chemotherapy and radiotherapy are based on excess RS production to kill cancer cells. With chemotherapy, the excess RS can cause oxidative stress in the microbiota when the drugs reach the host's gut. The changes in the oxidative stress are reflected in the "redox status," or more specifically, the redox ratio of the organisms. To recall, the redox ratio is the ratio of

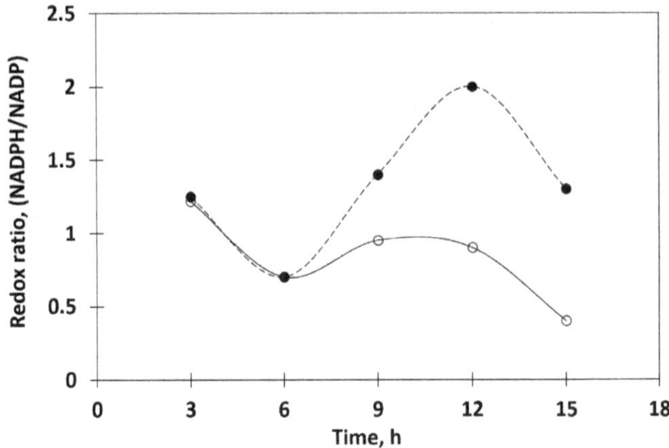

Figure 9.2.   Variation in the redox ratio (NADPH/NADP) of the whole cell with time when exposed to H_2O_2 (open circles) and in the control culture (filled circles)

NADPH to NADP levels in the cell, an important "metabolic state" indicator.

We studied *Enterococcus durans*, a representative gut bacterium that is nonpathogenic. We discovered [Jose *et al.*, 2018] that the cellular redox ratio oscillates with time under physiological conditions (Figure 9.2). Even more interestingly, the oxidative stress or RS decreased the redox ratio, as shown in Figure 9.2. The corresponding variation in hydroxyl radicals is shown in Figure 9.3. We showed through specific interactions [Jose *et al.*, 2018] that the hydroxyl radicals are significantly involved in this process.

We found an associated reduction in folate synthesized by the organism [Jose *et al.*, 2018]. Since the human host cannot synthesize folate, its production by the gut microbiome is significant — as mentioned earlier, folate plays a role in reducing the onset of colorectal cancer. Further, as expected, RS decreased the growth parameters of the gut bacterium. Thus, RS decreased redox ratio and folate secretion by the gut bacterium, which is directly detrimental to the human host. In addition, RS decreased the growth parameters of the gut bacterium, which is detrimental to the gut microbiome and, consequently, to the human host. Thus, one needs to be careful about the RS-induced and the resultant oxidative stress in the gut microbiome by anticancer drugs.

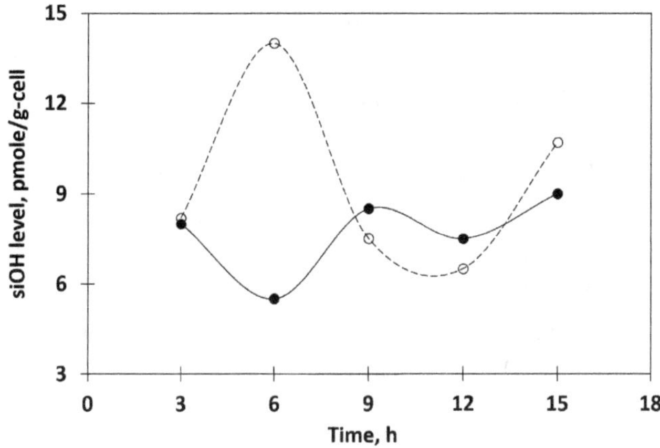

Figure 9.3.   Variation of specific intracellular hydroxyl radical (siOH) level with time in control (filled circles) and in cells exposed to H_2O_2 (open circles)

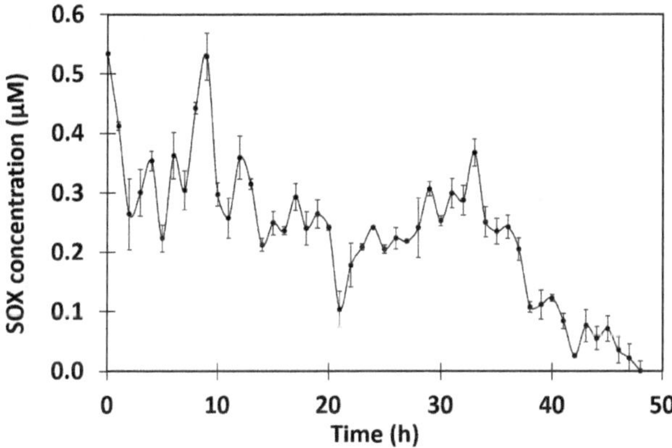

Figure 9.4.   A typical rhythm in superoxide concentration in HCT116 cells

9.3. Rhythms in RS in a Cancer Context

In Chapter 7, we discussed our discovery of RS rhythms (pseudo-steady-state [PSS] levels) in microalgae and its subsequent manipulation to significantly increase bio-oil yields. We also discovered the existence of RS rhythms in mammalian cells, especially cancer cells [Kizhuveetil *et al.*, 2019]. Our primary model system was the human colon cancer cell line, HCT116. An example of the rhythm in PSS concentrations of superoxide in HCT116 cells is shown in Figure 9.4.

The frequency of a typical rhythm under physiological conditions was about 0.042 h^{-1}. The time period was, thus, 23.8 h or almost 24 h. Thus, the superoxide rhythm under physiological conditions is circadian (approximately a day).

When the HCT116 cells are exposed to menadione (MD), a known superoxide generator, interestingly, the superoxide rhythm characteristics change (i.e., rhythm reset occurs). Figure 9.5 shows the variation in the time period of the rhythm as a function of MD concentration in a graphical form. The reduction in time period (T, in h) seems to be linear with MD concentration ([MD], in μM) as follows:

$$T = 23.52 - 1.05[MD]$$

Eq. 9.1

Interestingly, our studies with a p53–/– mutant cell line of HCT116 [Kizhuveetil *et al.*, 2019] showed that the ***change*** or ***reset*** in superoxide rhythm does not happen in the absence of the p53 protein. However, the ***endogenous*** superoxide rhythm is not dependent on the presence or absence of p53 — only the reset is dependent.

Molecular mechanism for the superoxide rhythm: We investigated the molecular mechanism for the rhythmic variation in intracellular superoxide levels [Kizhuveetil *et al.*, 2019] by studying the literature. We explored

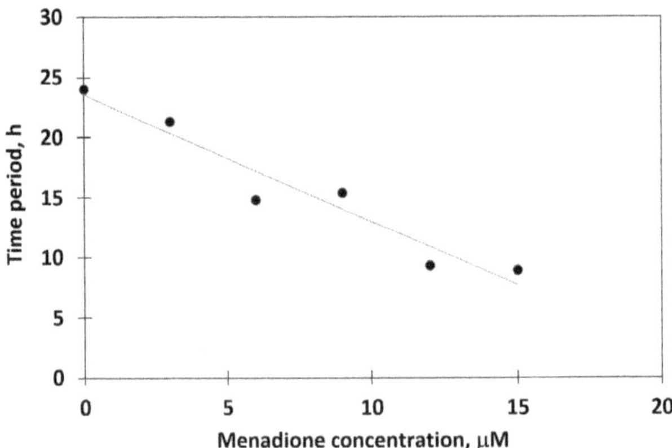

Figure 9.5.   Variation in the time period of the superoxide rhythm with exposure to different concentrations of menadione

the relevant pieces of the puzzle in the literature, which were published as disparate pieces of information, each complete in its own right. It is more accurate to say that we assembled the jigsaw puzzle with the relevant pieces of proven molecular information in the literature (for details, see Kizhuveetil *et al.*, [2019]). In brief,

- MD produces superoxide by redox cycling. The redox cycling consists of two steps, namely, conversion of MD into semiquinone (SQ) and reaction of SQ with oxygen to form superoxide.
- The superoxide influences the translocation of extracellular-signal-regulated kinase (ERK) from the cytoplasm into the nucleus.
- Inside the nucleus, ERK catalyzes the activation of the transcription factor, p53.
- The activated p53 binds to the promoter region of the *sod2* gene to produce the antioxidant enzyme superoxide dismutase (Mn-SOD). The ERK–p53–Mn-SOD link has also been established in the context of selenite-induced apoptosis in another cell line [Li *et al.*, 2010].
- Mn-SOD breaks down superoxide to hydrogen peroxide and oxygen. This "feedback loop" thus interferes with the translocation of ERK from the cytoplasm into the nucleus, which reduces the Mn-SOD formation. Therefore, there arises a time-based variation of intracellular superoxide levels with time, or in other words, a superoxide rhythm.

Thus, there is an involved molecular mechanism to control the superoxide level in the cell, which is represented schematically in Figure 9.6.

To understand the control aspects better, we built a mathematical model from the above molecular picture and associated aspects, which is described next.

Mathematical model: It is well known that a mathematical model is built for particular analysis and design purposes. It is also widely known that a successful mathematical model is only an abstraction of the important features of a system in the needed context and only some of the features. The experimental system could be highly complex, and a mathematical model is expected to capture only some of its aspects.

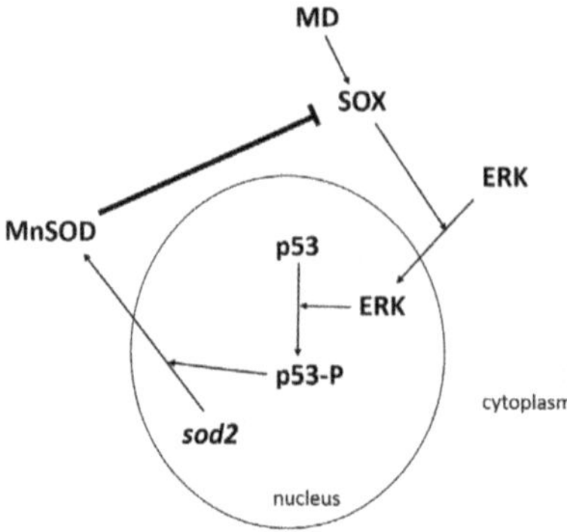

Figure 9.6.   The molecular mechanism for controlling superoxide (SOX) levels in the cell

The reactions considered for the mathematical model built for superoxide rhythm analysis are given next. The rate expression used for each reaction follows the reaction. The square brackets represent concentrations, as usual. In addition, in this model, SOX represents superoxide, $O_2^{\bullet-}$.

$$MD + NADPH \xrightarrow{\ CP450R\ } SQ + NADP^+ \qquad \text{Eq. 9.2}$$

$$r_1 = \frac{v_{m1}[MD][NADPH]}{K_N[MD] + K_M[NADPH] + [MD][NADPH]} \qquad \text{Eq. 9.3}$$

$$SQ + O_2 \underset{k_{m1}}{\overset{k_1}{\rightleftharpoons}} MD + SOX + H^+ \qquad \text{Eq. 9.4}$$

$$r_2 = k_1[O_2][SQ] - k_{m1}[MD][SOX] \qquad \text{Eq. 9.5}$$

$$ERK_c + SOX \xrightarrow{\ k_{r1}\ } ERK_n + \varnothing \qquad \text{Eq. 9.6}$$

The subscripts c and n refer to cytoplasm and nucleus, respectively. The symbol $\varnothing$ represents the metabolic pool.

$$r_3 = kt_1[ERK]_c[SOX] \qquad \text{Eq. 9.7}$$

$$P + p53 \underset{phosphatase}{\overset{ERK2}{\rightleftharpoons}} p53 - p \qquad \text{Eq. 9.8}$$

$$r_4 = \frac{k_3 n_v[ERK]_c[p53]}{k_{p53} + [p53]} - \frac{V_{max4}[p53 - P]}{K_{p53-P} + [p53 - P]} \qquad \text{Eq. 9.9}$$

$$2SOX + 2H^+ \xrightarrow{\ Mn-SOD\ } H_2O_2 + O_2 \qquad \text{Eq. 9.10}$$

$$r_5 = \frac{k_2[Mn-SOD][SOX]}{K_{SOX} + [SOX]} \qquad \text{Eq. 9.11}$$

$$\phi + p53 - P + DNA \rightarrow mSOD2 \qquad \text{Eq. 9.12}$$

$$r_6 = \frac{I_{max}[D_n][p53 - P]^{n_a}}{K_{p53-P}^{n_a} + [p53 - P]^{n_a}} \qquad \text{Eq. 9.13}$$

$$mSOD2 \rightarrow \phi \qquad \text{Eq. 9.14}$$

$$r_7 = k_{dm}[mSOD2] \qquad \text{Eq. 9.15}$$

$$\phi + mSOD2 \rightarrow Mn - SOD + mSOD2 \qquad \text{Eq. 9.16}$$

$$r_8 = k_{tr}[mSOD2] \qquad \text{Eq. 9.17}$$

$$Mn - SOD \rightarrow \phi \qquad \text{Eq. 9.18}$$

$$r_9 = k_{dp}[Mn - SOD] \qquad \text{Eq. 9.19}$$

$$H^+ + NADP^+ \xrightarrow{\ FNR\ } NADPH \qquad \text{Eq. 9.20}$$

$$r_{10} = \frac{V_{max5}[NADP^+]}{K_{NADP^+} + [NADP^+]} \qquad \text{Eq. 9.21}$$

Further, the rate of oscillation is given by

$$r_{oscil} = -LAe^{-Lt} + Ae^{-Lt}(wcos(wt + phi) - Lsin(wt + phi)) \qquad \text{Eq. 9.22}$$

Material balances for the relevant species in the above reactions in equations 9.2–9.20 are as follows:

$$\frac{d[MD]}{dt} = -r_1 + r_2 \qquad \text{Eq. 9.23}$$

$$\frac{d[NADH]}{dt} = -r_1 + r_{10} \qquad \text{Eq. 9.24}$$

$$\frac{d[NADP^+]}{dt} = r_1 - r_{10} \qquad \text{Eq. 9.25}$$

$$\frac{d[SQ]}{dt} = r_1 - r_2 \qquad \text{Eq. 9.26}$$

$$\frac{d[SOX]}{dt} = r_2 - r_3 - 2r_5 + r_{oscil} \qquad \text{Eq. 9.27}$$

The superoxide formation details are complex. The term for the rate of oscillation in Eq. 9.27 was included to address the endogenous levels of intracellular superoxide.

$$\frac{d[ERK]_c}{dt} = -r_3 \qquad \text{Eq. 9.28}$$

$$\frac{d[ERK]_n}{dt} = n_v r_3 \qquad \text{Eq. 9.29}$$

$$\frac{d[p53]}{dt} = -r_4 \qquad \text{Eq. 9.30}$$

$$\frac{d[p53-P]}{dt} = r_4 \qquad \text{Eq. 9.31}$$

$$\frac{d[mSOD2]}{dt} = r_6 - r_7 \qquad \text{Eq. 9.32}$$

$$\frac{d[Mn-SOD]}{dt} = r_8 - r_9 \qquad \text{Eq. 9.33}$$

More complete details are available in the open-access journal paper [Kizhuveetil *et al.*, 2019]. It can be seen from that paper that the model was satisfactorily able to predict the superoxide rhythms and their MD-induced reset (also see Figure 9.7). Further, the model was able to predict the responses of superoxide rhythms, with and without MD, in another cell line, HepG2, a hepatoma (liver cancer) cell line. The model was also able to predict the superoxide rhythms when another superoxide inducer, doxorubicin, was used instead of MD. Doxorubicin is a proven redox-cycling drug for cancer management.

9.4. The Importance of RS Rhythms in Cancer Treatment/Management

In cancer treatment, the dynamic changes in the intracellular RS levels are usually not considered. However, chemotherapy and radiotherapy are

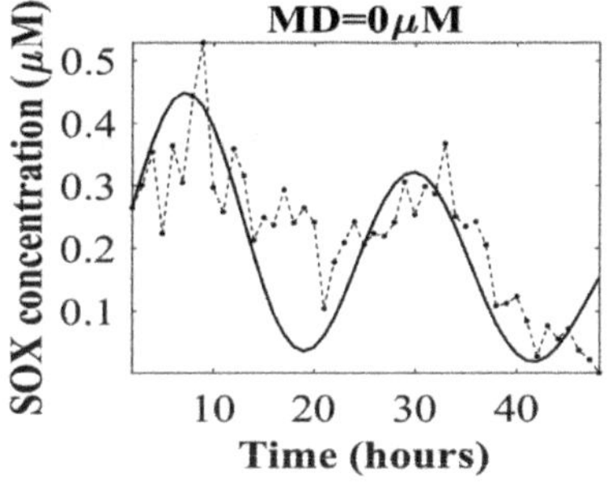

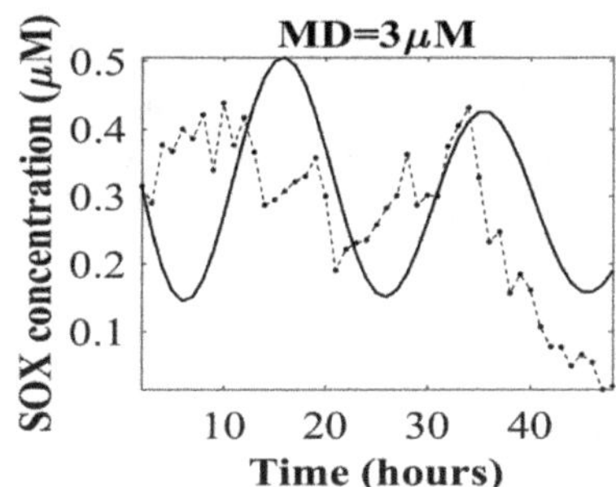

Figure 9.7.   Superoxide rhythm, native and that reset by 3 μM menadione. The points and dotted curves represent the experimental data. The solid lines represent the mathematical model predictions

based on inducing high RS levels in cancer cells to kill them. Ignoring the rhythms in intracellular RS levels and their entrainment may be a reason for undesirable situations, such as drastic side effects of chemotherapeutic drugs in unaffected cells or the ineffectiveness of the administered drug. Thus, planning chemotherapy strategies in conjunction with the RS rhythms is important.

We studied two cancer cell systems, a cervical cancer cell line, SiHa, and a colon cancer cell line, HCT116. We studied the effects of two RS-generating drugs/compounds, curcumin and MD, on the above cell lines. The detailed data are available in Kuzhuveetil *et al.*, [2020]. Let us look at the data highlights from the cancer treatment and management perspective. We exposed the cells to the RS-generating compounds, curcumin and MD, individually, at their IC_{50} levels at three different times from the start of the culture — that is, either at 0 h, 4 h, or 8 h. Upon exposure, the RS-generating compounds entrained (modified or reset) the RS rhythm characteristics. Tables 9.1 and 9.2 provide the data on the entrainment for SiHa and HCT116 cells, respectively. For example, in SiHa cells, curcumin reset the superoxide rhythm period from 15.4 h (endogenous rhythm) to 9.6 h (addition at 0 h), 9.2 h (addition at 4 h), or 12.1 h (addition at 8 h). In the same cells, curcumin reset the hydroxyl radical rhythm from 25 h (endogenous rhythm) to 11.1 (addition at 0 h), 8.3 h (addition at 4 h), 23.8 h (addition at 8 h), and so on. More importantly, the cytotoxicity of curcumin, when added at 4 h, was 27% higher than when added at 0 h. Thus, the drug can be at least 27% more effective

Table 9.1.   Entrainment in rhythm time period in SiHa cells

Reactive species	Drug addition time, h after start	Control, h	Curcumin, h	Menadione, h
Superoxide	0	15.4	9.6	11.0
	4		9.2	9.6
	8		12.1	12.4
Hydroxyl	0	25.0	11.1	21.7
	4		8.3	20.4
	8		23.8	25.0

Table 9.2.   Entrainment in rhythm time-period in HCT116 cells

Reactive species	Drug addition time, h after start	Control, h	Curcumin, h	Menadione, h
Superoxide	0	22.6	8.2	12.0
	4		9.2	12.8
	8		No rhythm	12.4
Hydroxyl	0	20.2	8.9	15.4
	4		9.6	20.3
	8		10.1	20.3

if added at the appropriate time in the intracellular RS rhythm. In other words, by a mere change in the administration time, the same drug at the same concentration can be much more effective against cancer.

The above results are valid under normoxic conditions (21% oxygen in the supplied air). Recent results from our laboratory seem to suggest that hypoxic conditions (say, 1% oxygen) could alter the RS rhythms and toxicity levels of a redox agent to the HCT116 cancer cell line. Further studies to understand this phenomenon better are in progress.

9.5. RS and Shear Stress in a Metastasis Context

In an earlier chapter, we saw that shear stress impacted the growth characteristics and productivity of *Bacillus subtilis*. Many years after that work, we wanted to explore whether shear stress can be used to kill or deactivate metastasizing cancer cells. Metastasis is the process by which the cancer cells migrate from the primary target organ/tissue to the secondary organ/tissue via the blood flow [Chaffer and Weinberg, 2011].

Although, in most cases, the primary cancer is cured by chemotherapy, radiation therapy, and/or surgery, more than 90% of cancer-related deaths are caused by metastasis [Chaffer and Weinberg, 2011]. The cancer cells are subject to shear stress during metastasis and related, dangerous, cancer-related events (cancer metastatic cascade) in the body.

Shear stress is known to play significant roles in the various steps of the cancer metastatic cascade [Wirtz *et al.*, 2011; Huang *et al.*, 2018]. Understandably, shear stress affects the cancer cells when they move through the vasculature in the fluid streams. However, shear stress also affects during pre-intravasation and post-extravasation. Pre-intravasation refers to the process by which the cancerous cells move through the surrounding tissues at their original location and enter the bloodstream. Post-extravasation refers to the process by which the cancerous cells move from the bloodstream to the new organ/location. In addition, shear affects the other processes in the metastasis cascade, such as adhesion and proliferation [Ma *et al.*, 2018].

We studied the effects of defined laminar shear on HCT116 colon cancer cells. We subjected the HCT116 cells to defined shear using a specially designed and fabricated cone-and-plate device [KrishnaPriya *et al.*, 2023]. We found a linear relationship between shear stress and RS (the molecular mediators of any cellular stress), especially the hydroxyl and superoxide radicals. This linear relationship was expected from our experience with shear effects on *B. subtilis*, which is detailed in Chapter 7.

When we began the work, we felt that the shear effects that are molecularly propagated through RS could potentially be transformed into a tool for cancer therapy through a better understanding of the shear-related micro RNA (miRNA) molecules. The miRNAs are about 18–25 nucleotides long; they are noncoding RNA capable of degrading or blocking the translation of the protein-coding mRNA [Cai *et al.*, 2009]. MicroRNAs play diverse roles in cellular metabolic processes regulating multiple mRNAs simultaneously, based on sequence complementarity

[Cai *et al.*, 2009]. MicroRNA controls the action of oncogenes and tumor suppressor genes; the imbalance in the action of oncogenes and tumor suppressor genes triggers carcinogenesis [Rupaimoole and Slack, 2017].

We identified and verified some microRNAs that can be explored to manipulate shear effects on colon cancer cells through various analyses of the data in the literature available in many databases and published papers. The identified miRNAs were hsa-miR-335-5p, hsa-miR-26b-5p, and hsa-miR-34a-5p. If interested in the details, please see KrishnaPriya *et al.* [2022].

The detailed experimental investigations on this system [KrishnaPriya *et al.*, 2023] showed that only the shear-induced, differential expressions of hsa-miR-335-5p and hsa-miR-34a-5p were statistically significant. Upregulation of has-miR-335-5p, but downregulation of has-miR-34a-5p were found. Thus, they need to be used appropriately in miRNA-based therapies to manage metastasis. Our suggestions on probable miRNA-based therapies and the relevant summary of our findings are presented in Figure 9.8.

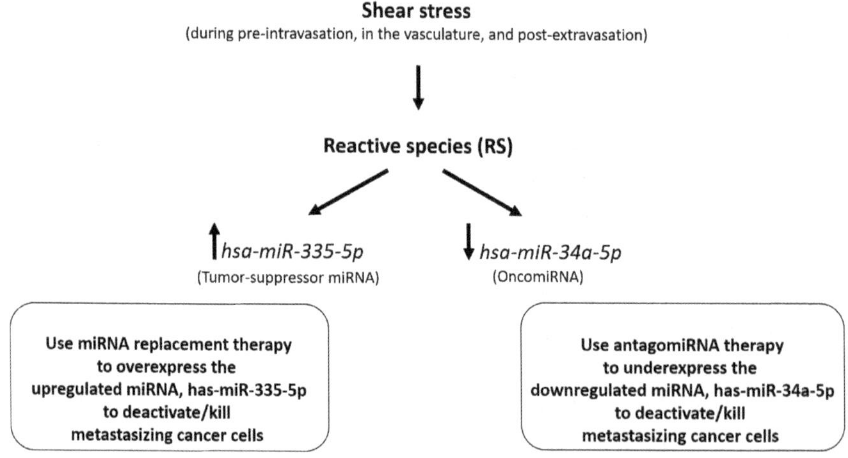

Figure 9.8. The proposed strategy to use the shear stress experienced by the metastasizing cancer cells to deactivate or kill them through appropriate microRNA therapies. Reactive species are the molecular mediators of shear stress effects in the cell

References

Cai, Y., Yu, X., Hu, S., and Yu, J. (2009). A brief review on the mechanisms of miRNA regulation. *Genom. Proteom. Bioinform.*, 7, 147.

Chaffer, C. L., and Weinberg, R. A. (2011). A perspective on cancer cell metastasis. *Science*, 331, 1559.

Galadari, S., Rahman, A., Pallichankandy, S., and Thayyullathil, F. (2017). Reactive oxygen species and cancer paradox: to promote or to suppress? *Free Rad. Biol. Med.*, 104, 144.

Huang, Q., Hu, X., He, W., Zhao, Y., Hao, S., Wu, Q., Li, S., Zhang, S., and Shi, M. (2018). Fluid shear stress and tumor metastasis. *Am. J. Cancer Res.*, 8, 763.

Jose, S., Bhalla, P., and Suraishkumar, G. K. (2018). Oxidative stress decreases the redox ratio and folate content in the gut microbe. *Enterococcus durans* (MTCC 3031), *Sci. Rep.*, 8, 12138.

Kizhuveetil, U., Palukuri, M. V., Sharma, P., Karunagaran, D., Rengaswamy, R., and Suraishkumar, G. K. (2019). Entrainment of superoxide rhythm by menadione in HCT116 colon cancer cells. *Sci. Rep.*, 9, 3347.

Kizhuveetil, U., Omer, S., Karunagaran, D., and Suraishkumar, G. K. (2020). Improved redox anti-cancer treatment efficacy through reactive species rhythm manipulation. *Sci. Rep.*, 10, 1588.

KrishnaPriya, S., Omer, S., Banerjee, S., Karunagaran, D., and Suraishkumar, G. K. (2022). An integrated approach to understand fluid shear stress-driven and reactive oxygen species-mediated metastasis of colon adenocarcinoma through mRNA-miRNA-lncRNA-circRNA networks. *Mol. Genet. Genom.*, 297, 1353.

KrishnaPriya, S., Nair, P. S., Bhalla, P., Karunagaran D., and Suraishkumar G. K. (2023). Shear stress and microRNAs: toward better metastatic cancer management. *Biotech. Prog.*, e3396.

Kim, E., Jang, M., Song, M., Kim, D., Kim, Y., and Jang, H. H. (2019). Redox-mediated mechanism of chemoresistance in cancer cells. *Antioxidants*, 8, 471.

Li, Z., Shi, K., Guan, L., Cao, T., Jiang, Q., Yang, Y., and Xui, C. (2010). ROS leads to MnSOD upregulation through ERK2 translocation and p53 activation in selenite-induced apoptosis of NB4 cells. *FEBS Lett.*, 584, 2291.

Ma, S., Fu, A., Chiew, G. G. Y., and Luo, K. Q. (2017). Hemodynamic shear stress stimulates migration and extravasation of tumor cells by elevating cellular oxidative level. *Cancer Lett.*, 388, 239.

Rupaimoole, R., and Slack, F. J. (2017). MicroRNA therapeutics: towards a new era for the management of cancer and other diseases. *Nat. Rev. Drug Discov.*, 16, 203.

Wirtz, D., Konstantopoulos, K., and Searson, P. C. (2011). The physics of cancer: the role of physical interactions and mechanical forces in metastasis. *Nat. Rev. Cancer*, 11, 512.

Chapter 10

A Reactive Species Module for Integration into Metabolic Networks

In the previous chapters, we saw that reactive species (RS) play significant roles in cell signaling but are toxic at high levels. RS are now reasonably accepted as the fundamental molecular mediators of stress effects on cells and are, thus, important in many disease contexts. However, a comprehensive, in-depth understanding of the various effects of RS on cells is rudimentary because the metabolic networks are complex.

We have developed a *scalable* metabolic RS module that can be integrated with potentially *any* metabolic model or any genome-scale metabolic (GSM) model to gain insights into RS effects. The RS module is also expected to significantly contribute to the comprehensive, in-depth understanding of RS effects on cells.

10.1. Genome-Scale Metabolic Models (With Archanaa Sundararaghavan)

First, let us look at what are GSM models. GSM models are standard analysis tools to gain a systems-level understanding of cellular processes. GSM models consist of a reasonably comprehensive set of reactions occurring in the cell and a representation of the genes that code for the enzymes that catalyze the relevant metabolic reactions. GSM models typically capture the organism-specific biochemical transformations, their participating metabolites, and genes that encode the enzymes catalyzing the transformations. GSM models of various organisms and cell lines have been built and used for many biotechnological purposes. Examples include organisms such as *Escherichia coli*, *Saccharomyces cerevisiae*, and so on [Zhang and Hua, 2016], and Chinese hamster ovary cell lines [Calmels *et al.*, 2019].

GSM models serve as excellent platforms to infer genotype–phenotype relationships. They are helpful tools for exploring the mechanistic basis of various cellular responses subject to various environmental/nutrient conditions [Palsson, 2015]. For example, Sundararaghavan *et al.* [2020] resorted to GSM to gain insights into the mechanism behind oxidative stress-induced increase in TAG accumulation of oleaginous organisms. The authors, from the literature, identified two different targets for RS that are relevant to the TAG accumulation pathway based on which two mechanisms underlying the oxidative stress-mediated TAG increase were hypothesized (Figure 10.1). The hypothesized mechanism (HM) details are as follows:

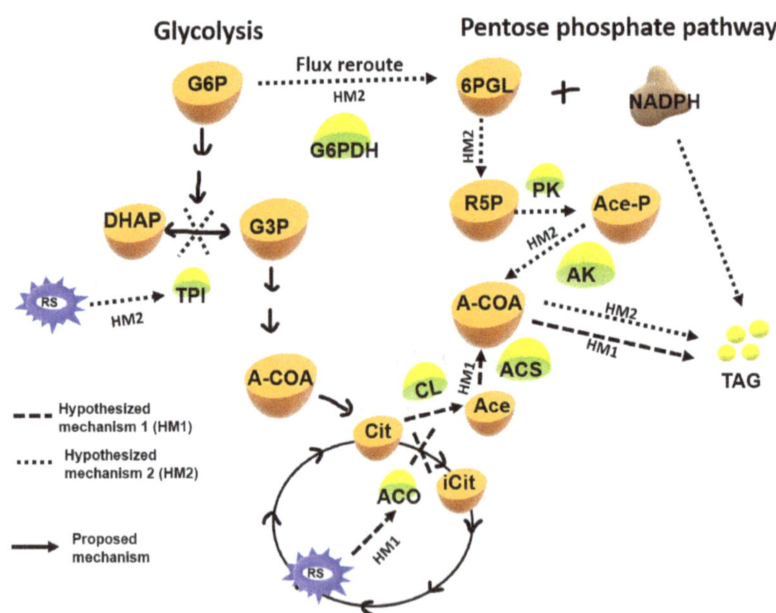

Figure 10.1. Schematic representation of possible mechanisms underlying oxidative stress-mediated increase in TAG accumulation of oleaginous organisms

HM1. RS inhibits Aconitase (TCA-cycle enzyme) followed by rerouting of acetyl-coA (TAG precursor) toward TAG synthesis.

HM2. RS inhibits Triosephosphate isomerase (TPI, glycolytic enzyme) followed by activation of NADPH (TAG co-factor) sourcing pentose pyruvate pathway (PPP), which in turn activates flux through phosphoketolase (PK) and enhances the flux through phosphotransacetylase (PTA) followed by increased acetyl-coA synthesis and eventual TAG synthesis.

The authors tested the hypothesis using the reconstructed GSM model of an oleaginous bacterium, *Rhodococcus opacus* PD630. The oxidative stress was simulated in the model by inhibiting the flux through the respective RS targets. The HM1 was nullified as inhibiting the aconitase flux did not improve the TAG flux (Table 10.1).

The experimental observation of poorly correlated aconitase activity and acetyl-CoA levels also supported the model prediction. Upon testing HM2, the authors observed that the inhibition of TPI flux activated the PPP flux, and as the inhibition of TPI flux increased, the PPP flux increased concomitantly. The increased PPP flux eventually resulted in increased TAG flux (Table 10.1). The model prediction was further supported by experimental observation of decreased TPI activity and increased glucose-6-phosphate dehydrogenase (G6PDH) activity (PPP enzyme) and TAG levels in *R. opacus* PD630 under oxidative stress. Thus, with the help of the GSM and experimental evidence, HM2 was proposed

Table 10.1.   Hypothesis testing using GSM of *R. opacus* PD630

Conditions	TAG flux (mmol (g cell)$^{-1}$ h^{-1})	Increase in TAG flux (%)
Active aconitase and TPI flux (control)	0.022	–
Inhibited aconitase flux (HM1)	0.021	0
Inhibited TPI flux (HM2)	0.053	153

as a possible mechanism underlying the oxidative stress-mediated increase in TAG accumulation of oleaginous organisms. The authors also made another significant observation from the model, which is as follows: rerouted or enhanced flux through PPP could be essential to improve the TAG levels of oleaginous organisms as it can account for a simultaneous supply of acetyl-CoA and NADPH that is vital for TAG synthesis.

In addition, a GSM model was used to gain good insights during the bio-cementation process. The specific aspect was the effect of medium C/N ratio on the process. Details can be found in Murugan *et al.* [2022].

10.2. RS Module Development

At the time of this writing, the genome-scale models (metabolic reconstructions) do not include reactions of RS, except for a few ROS detoxification reactions. Therefore, about six years ago, we embarked on a journey to develop an RS module that was as comprehensive as possible. We did not know the journey would take this long—I expected it to happen in a year or two. This was after many years of effort through kinetic equations, after which we finally accepted that such an approach, although rigorous, would not scale due to the nonavailability of kinetic constants.

The RS module development consisted of three broad steps [Sridhar *et al.*, 2023].

1. We consulted various databases, such as NIST, NRDL, Pubchem, VMH, and so on, to get the basic reactions. The reactions were manually curated, which took much time.
2. Then, we completed a large number of incomplete reactions, verified mass balances, verified charge balances, and removed thermodynamically infeasible cycles (TIC) that also arose due to improperly included reversible reactions, and so on. We used several available programs, such as Chemsketch, Marvin, CycleFreeFlux for TIC, and so on, and arrived at the improved RS module.
3. To incorporate the improved RS module into the needed GSM model on the COBRA platform, we used programs such as rBioNet and many COBRA toolbox functions (mergeTwoModels, verifyModels, etc.).

The RS module is available at:
https://drive.google.com/drive/folders/1sKsS9RoeZ3kflMC874_-5_
-d62q045_2

10.3. Application of the RS Module (With Subasree Sridhar)

In this section, let us see an application of the RS module to gain improved insights into cancer metabolism.

10.3.1. *Introduction to metabolic modeling of cancer*

Cancer is a metabolic disease because metabolic reprogramming/rewiring is a critical hallmark of cancers [Hanahan and Weinberg, 2011]. The availability of a huge amount of omics data for cancerous tissues and models such as cell lines and animal models has helped in understanding the metabolic dysregulation (changes in the cellular reaction rates/fluxes) in cancers through the use of GSM models [Agren *et al.*, 2014; Turanli *et al.*, 2019].

To make better sense of the omics data, we need mechanistic models. The mechanistic models help us better understand the metabolic rewiring in cancers. The two broad categories of the available modelling frameworks to study cancer are kinetic models and constraint-based models (GSM models) [Volkova *et al.*, 2020]. The main limitation of using kinetic models is the nonavailability of the needed kinetic parameters. The determination of the kinetic parameters for one set of conditions is itself a tedious process. On the other hand, constraint-based models that employ GSM models are easy to formulate because they rely only on the stoichiometrically balanced biochemical reactions of a system.

Another crucial advantage of using GSM models is that many types of omics data can be integrated with them [Yizhak *et al.*, 2015]. Thus, integrating the omics data with the GSM models seems to be the best way to understand the metabolic dysregulation occurring in cancers. However, unlike kinetic models, GSM models, by themselves, cannot represent the dynamics of the system because they can only describe the steady-state features. The output of the GSM models is usually described in terms of steady-state fluxes (rates) of reactions, which have been normalized with the cell mass, mmol $(gDW)^{-1}$ h^{-1}. The steady state implies that the flux of a metabolite remains constant with time, and hence, there is no accumulation of the metabolite.

Large-scale, human whole-body genome reconstructions such as Recon 3D and Human-GEM are used as the template models from which context-specific models are built [Brunk *et al.*, 2018; Robinson *et al.*, 2020]. The available data are used to extract context-specific GSM models from the whole-body genome reconstructions. For example, colorectal cancer (CRC) GSM models are built by integrating colorectal tumor data onto the whole body genome reconstructions. GSM models are steady-state models that are optimized for an objective function(s) through linear programming [Feist and Palsson, 2010]. Different types of constraints, such as environmental and thermodynamic constraints, are applied to arrive at the optimal solution by reducing the solution space (Orth *et al.*, 2010). Cancer cells strive to proliferate, which justifies using the biomass growth equation as the objective function for cancer metabolic models. The biomass growth equation is an equation that constitutes the consumption of biomolecules along with ATP hydrolysis toward biomass formation. An example of a set of environmental constraints that are imposed on the GSM models is the set of uptake rates of nutrients from DMEM-high glucose medium [Bhalla *et al.*, 2022; Sridhar *et al.*, 2023].

10.3.2. *Modelling RS in cancer metabolic models*

As mentioned earlier, there is a dearth of information on RS in the two whole-human models of Recon 3D and Human-GEM. Therefore, we thought of developing a RS module containing RS reactions that can complement the other extensive set of metabolic reactions. The RS are closely associated with the development of cancers. Thus, the inclusion of RS metabolic reactions in the cancer GSMs is vital to get a more precise picture of the cancer metabolism. It may be interesting to note that many RS reactions are spontaneous, and they occur without any involvement of genes/enzymes.

We have created an RS module of metabolic reactions associated with RS such as reactive oxygen species (ROS), reactive nitrogen species (RNS), reactive sulfur species (RSS), reactive halogen species (RHS), and reactive carbonyl species (RCS) based on information from the literature and databases [e.g., Kruk and Aboul-Enein, 2017; Marcolongo *et al.*, 2019; https://kinetics.nist.gov/solution/]. These reactions are relevant to human metabolism, and the RS reaction is scalable — they can be integrated with any human GSM model wherever RS is involved, which is everywhere because RS are a part of the basal metabolism at homeostasis. The RS reaction module is designed for Recon 3D-derived GSM models, but can be made suitable for any other GSM models by changing the nomenclature of the metabolites, which is not a highly time-consuming job. To illustrate its application, we have integrated this RS reactions module to understand their contributions and lead to improved in-silico predictions of three cancer metabolic models, namely, triple negative breast cancer (TNBC), high-grade serous ovarian cancer (HGSOC), and CRC. We have integrated the RS reactions module with three cancer cell line models, HCC1143 (TNBC), OVKATE (HGSOC) and SW620 (CRC). The cell lines were chosen based on their similarity to the molecular phenotype of the respective tumors [Grigoriadis *et al.*, 2012; Mitra *et al.*, 2015; Ronen *et al.*, 2019].

The cell line metabolic models are constructed using Recon 3D as the template model [Brunk *et al.*, 2018] and transcriptomics data from CCLE database [Ghandi *et al.*, 2019]. The SWIFTCORE algorithm from Cobra Toolbox was used to integrate the transcriptomics data onto the Recon 3D model using the core reactions of each of the cancers as the input, which was obtained using the local T2 algorithm [Tefagh and Boyd, 2020]. The requirements for building a context specific model are given in Figure 10.2.

The developed RS reactions module was integrated into the three cancer models using the mergeTwoModels algorithm from Cobra Toolbox, where model 1 is any context-specific model with a defined objective function and model 2 is the RS reactions module. The integration of the RS reactions module to a GSM model is represented in Figure 10.3.

The RS reactions module integrated and unintegrated standard cancer models for the three cancers are named as follows:

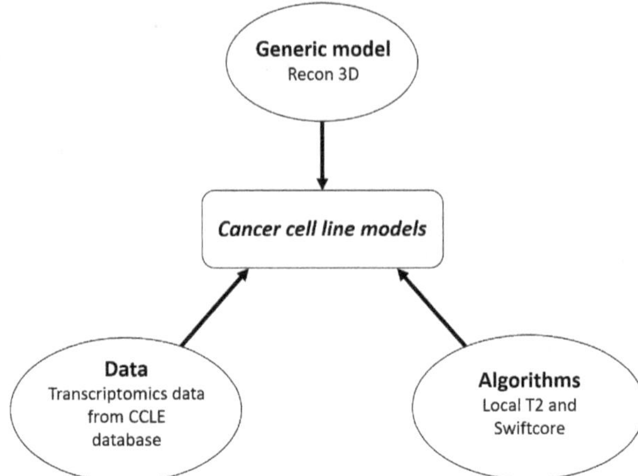

Figure 10.2. The requirements for building the three cancer cell line models

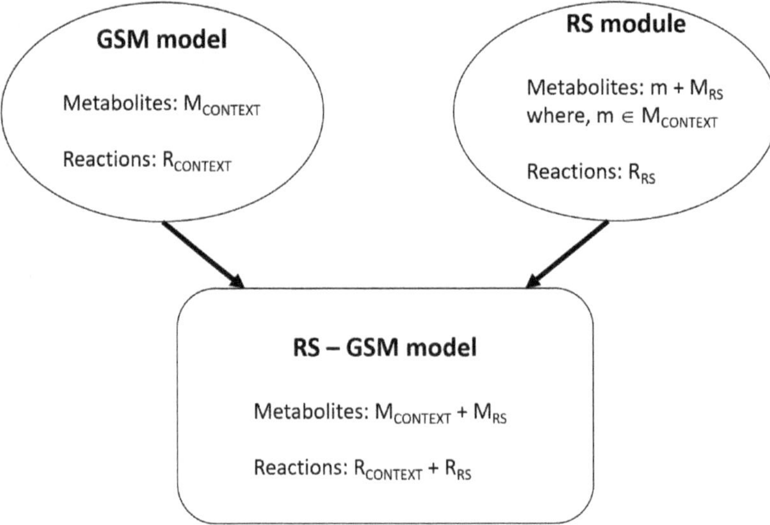

Figure 10.3. Integration of the RS module with a GSM model

RS-TNBC model and TNBC model for HCC1143 model
RS-HGSOC and HGSOC model for OVKATE model, and
RS-CRC and CRC model for SW620 model.

The fluxes of many reactions in the RS-cancer GSM models differed from the cancer GSM models. This change in the fluxes is due to the interaction of the RS reactions with the cancer GSM model reactions. The RS reactions module shares some metabolites with the cancer GSM models, through which the interaction occurs. Some common metabolites are ATP, proton, NAD(P), NAD(P)H, glutathione, dihydroascorbate, and others.

The set of reaction fluxes obtained through the constrained optimization is not unique. In other words, a distribution in the set of fluxes in the solution space exists that satisfies the optimization constraints. A better acceptable set of reaction fluxes in this context can be obtained through flux sampling analysis. Flux sampling analysis gives a chain of feasible fluxes of reactions, that is, probability distribution of reaction fluxes of the GSM models under the given constraints [Herrmann *et al.*, 2019]. The distribution of flux samples from the GSM models is generally skewed, due to which the median values are selected over the mean for analysis. The median of the 50,000 reaction fluxes was calculated for the cancer GSM models with and without RS reactions module integration. To illustrate the importance of integrating the RS module into GSM models, we have chosen the major metabolic reactions that are the hallmarks of each of the three cancers. The flux differences of these reactions are shown in RS-cancer and their respective cancer GSM models. The median of the 50,000 flux samples of these hallmark reactions is compared in the RS-cancer and the cancer GSM models, and they are mentioned in the following subsection. Our publication discusses them in detail — a details-oriented reader is directed to Sridhar *et al.* [2023].

10.3.3. *Better highlights of cancer metabolism from RS-cancer models*

The following are some of the molecular aspects that were better highlighted in RS-cancer models:

TNBC

- Increased dependence on glutamine metabolism for their cellular needs (glutamine addiction)
- The increased uptake of the glutamine through glutamine transporters such as SLC38A2, SLC7A5, and SLC1A5

HGSOC

- Aberration in polyamine metabolism
- Aberration in sphingolipid metabolism

CRC

- Increased glycolysis
- Increased fatty acid synthesis

Further, the differences in the regulation of ferroptosis were better highlighted in the three cancer GSM models upon RS module integration. Ferroptosis is a recently recognized form of programmed cell death that is caused by metabolic activities of the cancer cells, such as iron accumulation and lipid peroxidation [Stockwell *et al.*, 2017]. The three important metabolic pathways that decide the fate of ferroptosis-induced cell death are fatty acid or lipid metabolism, iron metabolism, and arachidonic acid metabolism [Stockwell *et al.*, 2017].

The RS-integrated models highlighted the differences in the metabolic reactions in the ferroptosis regulation in the three cancer models. This can be exploited in appropriately targeting the individual cancers for ferroptosis.

References

Agren, R., Mardinoglu, A., Asplund, A., Kampf, C., Uhlen, M., and Nielsen, J. (2014). Identification of anticancer drugs for hepatocellular carcinoma through personalized genome-scale metabolic modeling. *Mol. Syst. Biol.*, 10, 721.

Bhalla, P., Rengaswamy, R., Karunagaran, D., Suraishkumar, G. K., and Sahoo, S. (2022). Metabolic modeling of host–microbe interactions for therapeutics in colorectal cancer. NPJ Syst. Biol. Appl., 8, 1.

Brunk, E., Sahoo, S., Zielinski, D. C., Altunkaya, A., Drager, A., Mih, N., Gatto, F., Nilsson, A., Preciat Gonzalez, G. A., Aurich, M. K., Prlić, A., Sastry, A., Danielsdottir, A. D., Heinken, A., Noronha, A., Rose, P. W., Burley, S. K., Fleming, R. M. T., Nielsen, J., Thiele, I., and Palsson, B. O. (2018). RECON3D enables a three dimensional view of gene variation in human metabolism. *Nat. Biotechnol.*, 36, 272.

Calmels, C., McCann, A., Malphettes, L., and Andersen, M. R. (2019). Application of a curated genome-scale metabolic model of CHO DG44 to an industrial fed-batch process. *Met. Eng.*, 51, 9.

Feist, A. M., and Palsson, B. O. (2010). The biomass objective function. *Curr. Opin. Microbiol.*, 13, 344.

Ghandi, M., Huang, F. W., Jane-Valbuena, J., Kryukov, G. V., Lo, C. C., McDonald III, E. R., Barretina, J., Gelfand, E. T., Bielski, C. M., Li, H., *et al.* (2019). Next-generation characterization of the cancer cell line encyclopedia. *Nature*, 569, 503.

Grigoriadis, A., Mackay, A., Noel, E., Wu, P. J., Natrajan, R., Frankum, J., Reis-Filho, J. S., and Tutt, A. (2012). Molecular characterisation of cell line models for triple-negative breast cancers. *BMC Genom.*, 13, 1.

Hanahan, D., and Weinberg, R. A. (2011). Hallmarks of cancer: the next generation. *Cell*, 144, 646.

Herrmann, H. A., Dyson, B. C., Vass, L., Johnson, G. N., and Schwartz, J.-M. (2019). Flux sampling is a powerful tool to study metabolism under changing environmental conditions. *NPJ Syst. Biol. Appl.*, 5, 1.

Kruk, J. Y., and Aboul-Enein, H. (2017). Reactive oxygen and nitrogen species in carcinogenesis: implications of oxidative stress on the progression and development of several cancer types. *Mini Rev. Med. Chem.*, 17, 904.

Marcolongo, J. P., Venancio, M. F., Rocha, W. R., Doctorovich, F., and Olabe, J. A. (2019). NO/H_2S "crosstalk" reactions. The role of thionitrites (SNO^-) and perthionitrites ($SSNO^-$). *Inorg. Chem.*, 58, 14981.

Mitra, A. K., Davis, D. A., Tomar, S., Roy, L., Gurler, H., Xie, J., Lantvit, D. D., Cardenas, H., Fang, F., Liu, Y., Loughran, E., Yang, J., Stack, M. S., Emerson, R. E., Dahl, K. D. C., Barbolina, M., Nephew, K. P., Matei, D., and Burdette, J. E. (2015). In vivo tumor growth of high-grade serous ovarian cancer cell lines. *Gynecol. Oncol.*, 138, 372.

Murugan, R., Sundararaghavan, A., Dhami, N. K., Mukherjee, A., and Suraishkumar, G. K. (2022). Importance of C/N ratio in microbial cement production: insights through experiments and genome-scale metabolic modelling. *Biochem. Eng. J.*, 186, 108573.

Orth, J. D., Thiele, I., and Palsson, B. O. (2010). What is flux balance analysis? *Nat. Biotechnol.*, 28, 245.

Palsson, B. O. (2015). *Systems biology: Constraint-based reconstruction and analysis* (2nd ed.), Cambridge University Press, Cambridge.

Robinson, J. L., Kocabaş, S. P., Wang, H., Cholley, P.-E., Cook, D., Nilsson, A., Anton, M., Ferreira, R., Domenzain, I., Billa, V., Limeta, A., Hedin, A., Gustafsson, J., Kerkhoven, E. J., Svensson, L. T., Palsson, B. O., Mardinoglu, A.,

Hansson, L., Uhlén, M., and Nielsen, J. (2020). An atlas of human metabolism. *Sci. Signal.*, 13, eaaz1482.

Ronen, J., Hayat, S., and Akalin, A. (2019). Evaluation of colorectal cancer subtypes and cell lines using deep learning. *Life Sci. Allian.*, 2, e201900517.

Sridhar, S., Bhalla, P., Kullu, J., Veerapaneni, S., Sahoo, S., Bhatt, N., and Suraishkumar, G. K. (2023). A reactive species reactions module for integration into genome-scale metabolic models for improved insights: application to cancer. *Metabol. Eng.*, 80, 78.

Stockwell, B. R., Angeli, J. P. F., Bayir, H., Bush, A. I., Conrad, M., Dixon, S. J., Fulda, S., Gascon, S., Hatzios, S. K., Kagan, V. E., *et al.* (2017). Ferroptosis: a regulated cell death nexus linking metabolism, redox biology, and disease. *Cell*, 171, 273.

Sundararaghavan, A., Mukherjee, A., Sahoo, S., and Suraishkumar, G. K. (2020). Mechanism of the oxidative stress-mediated increase in lipid accumulation by the bacterium, *R. opacus* PD630: experimental analysis and genome-scale metabolic modelling. *Biotechnol. Bioeng.*, 117, 1779.

Tefagh, M., and Boyd, S. P. (2020). Swiftcore: a tool for the context-specific reconstruction of genome-scale metabolic networks. *BMC Bioinfo.*, 21, 1.

Turanli, B., Zhang, C., Kim, W., Benfeitas, R., Uhlen, M., Arga, K. Y., and Mardinoglu, A. (2019). Discovery of therapeutic agents for prostate cancer using genome-scale metabolic modeling and drug repositioning. *EBioMed.*, 42, 386.

Volkova, S., Matos, M. R., Mattanovich, M., and Marin de Mas, I. (2020). Metabolic modelling as a framework for metabolomics data integration and analysis. *Metabolites*, 10, 303.

Yizhak, K., Chaneton, B., Gottlieb, E., and Ruppin, E. (2015). Modeling cancer metabolism on a genome scale. *Mol. Syst. Biol.*, 11, 817.

Zhang, C., and Hua, Q. (2016). Applications of genome-scale metabolic models in biotechnology and systems medicine. *Front. Physiol.*, 6, 1.

Index